ISO 9000
for
Software
Developers

Also available from ASQC Quality Press

Documenting Quality for ISO 9000 and Other Industry Standards
Gary E. MacLean

ISO 9000: Preparing for Registration
James L. Lamprecht

Software Quality Assurance and Evaluation
James H. Dobbins

Principles and Practices of TQM
Thomas J. Cartin

Root Cause Analysis: A Tool for Total Quality Management
Paul F. Wilson, Larry D. Dell, and Gaylord F. Anderson

To request a complimentary catalog of publications, call 800-248-1946.

ISO 9000 for Software Developers

Charles H. Schmauch

ASQC Quality Press
Milwaukee, Wisconsin

ISO 9000 for Software Developers
Charles H. Schmauch

Library of Congress Cataloging-in-Publication Data
Schmauch, Charles H.
    ISO 9000 for software developers / Charles H. Schmauch.
       p.   cm.
    Includes bibliographical references and index.
    ISBN 0-87389-246-1 (alk. paper)
    1. Computer software—Quality control.   I. Title.
    QA76.76.Q35S37   1994
    658.5'62'0240904—dc20   93-40710
                                                          CIP

10 9 8 7 6 5 4 3 2 1

ISBN 0-87389-246-1

Acquisitions Editor: Susan Westergard
Project Editor: Kelley Cardinal
Production Editor: Annette Wall
Marketing Administrator: Mark Olson
Set in Berkeley and Stone Sans by Montgomery Media, Inc.
Cover design by Montgomery Media, Inc.
Printed and bound by BookCrafters, Inc.

ASQC Mission: To facilitate continuous improvement and increase customer satisfaction by identifying, communicating, and promoting the use of quality principles, concepts, and technologies; and thereby be recognized throughout the world as the leading authority on, and champion for, quality.

For a free copy of the ASQC Quality Press Publications Catalog, including ASQC membership information, call 800-248-1946.

Printed in the United States of America.

  Printed on acid-free recycled paper

ASQC
Quality Press
611 East Wisconsin Avenue
Milwaukee, Wisconsin 53202

Contents

Preface

This brief, easy-to-understand book provides an introduction to ISO 9000 for anyone who is interested in ISO 9000 or who has a need to know. Although it is beneficial to all readers, the book is particularly aimed at software developers and is written in language and terminology that should be familiar to most software developers.Written at an introductory level, the book is intended to provide a basic understanding of the ISO 9000 series of standards and insights into what is required to achieve ISO 9000 conformance and registration for the software development environment.

The ideas presented here are suggestions of ways to achieve ISO 9000 registration for the software development environment. The purpose of the book is to (1) provide a basic understanding of the ISO 9000 standards, their benefits, and how they apply to software development, (2) suggest what is needed in a software development quality system to satisfy the ISO 9000 standards, and (3) offer guidance on how to proceed to ISO 9000 registration. The book does not prescribe in detail how to implement an ISO 9000 conforming quality system; it leaves that to the developer. The best use of this book is to apply the information and ideas to your particular situation.

Whether you produce software for use outside of your organization, for in-house use, or for inclusion in some other product, and whether you are interested in ISO 9000 registration or not, if you develop software, you *should* be interested in ISO 9000. The ISO 9000 series of standards provides an excellent foundation with proven quality management principles and practices upon which to build your quality systems or against which to measure your existing development systems. The ISO 9000 series of standards provides a base upon which to build leading-edge quality for the products or service that you provide. It is a catalyst to quality. ISO 9000 conforming quality systems consistently yield products that meet customer expectations—and that is important to all software developers and their customers.

Beyond what is stated above, this book makes little attempt to justify ISO 9000 or convince readers that ISO 9000 is for them. Neither does the book intend to address total quality management or quality beyond that inherent with ISO 9000. The focus of this book is on ISO 9000 for software development—what it means, what it implies, and what is required to satisfy the ISO 9000 series of standards.

1

Introduction to ISO 9000

A Short Course on ISO 9000

Managing Quality

The objective of quality management is to produce quality products by building quality into the products rather than testing quality into the products. Quality management is meant to ensure that faults do not occur in the first place. Quality management systems[1] are used for developing products and are designed to ensure that quality is being designed and built into the products.

About the ISO 9000 Series of Standards

The ISO 9000 series of standards (interchangeably referred to as ISO 9000 throughout this book) is a series of international quality standards, developed by the International Organization for Standardization (see Figures 1.1 and 1.2), that applies to the quality management system and the process used to produce a product. ISO 9000 establishes a basic set of quality system requirements necessary to ensure that your process is capable of consistently producing products that meet the expectation of your customers. It provides an excellent base upon which to extend and improve your process, thereby improving the quality of your product, or service. ISO 9000 does not provide for leading-edge quality but does provide a strong quality foundation upon which you can build.

The ISO 9000 series of standards was originally developed for two-party contractual situations, mainly for the manufacturing environment. It is not written to any specific industry and is intended to be relevant to all types of businesses. Diverse industries, such as electronics, chemistry, automobile, transportation, health care, and banking, are implementing these standards.

Although many countries have established their own standards for quality assurance (see Figure 1.3), ISO 9000 has been accepted almost universally for quality assurance and country standards are reflecting ISO 9000. In the United States, for example, ANSI/ASQC Q90–Q94 are almost ISO 9000 verbatim. As of January 1993, 56 nations and the European Economic Community had adopted ISO 9000 (see Appendix J).

The International Organization for Standardization is a nongovern-mental organization that was established in 1946 to develop worldwide standards to improve international communication and to promote smooth and equitable growth of international trade. Headquartered in Geneva, Switzerland, the organization is currently composed of 91 member countries. The United States is represented by the American National Standards Institute (ANSI).

The International Organization for Standardization work results in international technical agreements that are published as international standards. All standards developed by the organization are voluntary, no legal requirements force countries to adopt them. However, countries and industries often adopt and attach further significance to the organization's standards; thereby, making the standards mandatory. In many respects, this is what is happening to the ISO 9000 series of standards.

Figure 1.1. A brief history of the International Organization for Standardization.

As markets become more open, manufacturers rely on their products being acceptable worldwide. A key to acceptance and success in a global marketplace will be higher quality. At the same time, purchasers have a grow-ing expectation that their suppliers can provide products that meet their needs and expectations. Further, customers are beginning to demand that purchased goods and services be the result of a process that includes a demonstrably effective quality management system. In most cases, ISO 9000 provides the level of assurance required by customers that their suppliers can fulfill their expectations.

The European Community (EC) has adopted quality management as a key strategic element in its drive to improve the competitiveness of European suppliers, not only in Europe but worldwide, and the EC has chosen ISO 9000 as its standards for quality assurance. Developing countries see informa-tion technology as a gateway to wealth and view ISO 9000 as providing a foothold into this business.

Prior to 1979, as quality was rapidly emerging as a new emphasis in commerce and industry, numerous national and multinational standards were developed for quality systems. These standards, although common in some areas, were not sufficiently consistent for use as international standards. In 1979, the organization formed Technical Committee 176 (TC176) to formulate an international standard for quality systems. In 1987, the ISO 9000 series of standards was published and has been adopted by many nations and is rapidly supplanting prior national and industry-based standards.

The adoption of ISO 9000 by the European Community (EC) has global significance because it places new marketplace pressures on all producers worldwide that wish to trade with European countries or even compete with European companies in other markets.

Figure 1.2. History of ISO 9000.

International	ISO 9001	ISO 9002	ISO 9003
EC	EN 29001	EN 29002	EN 29003
Belgium	NBN X 50-003	NBN X 50-004	NBN X 50-005
France	NF X 50-131	NF X 50-132	NF X 50-133
Germany	DIN ISO 9001	DIN ISO 9002	DIN ISO 9003
Netherlands	NEN 2646	NEN 2647	NEN 2648
Norway	NS 5801	NS 5802	NS 5803
Switzerland	SN 029 100A	SN 029 100B	SN 029 100C
United Kingdom	BS 5750:Part 1	BS 5750:Part 2	BS 5750:Part 3
United States	ANSI/ASQC Q91	ANSI/ASQC Q92	ANSI/ASQC Q93

Figure 1.3. Standards by country.

ISO 9000 is well established in the EC where over 20,000 organizations are registered to ISO 9000 standards. Non-ISO 9000 manufacturers will be finding it more and more difficult to compete in the European and worldwide market. They will be seeing stiffer competition from ISO 9000 registered suppliers. They will also begin to see more demanding customers. Manufacturers will need ISO 9000 conforming quality management systems to sustain their competitiveness in these markets. Because the standards require the supplier to ensure the quality not only of their final product, but of all the product's constituent parts regardless of origin, suppliers who provide parts to ISO 9000 manufacturers will also be seeing more stringent quality requirements from their customers. ISO 9000 manufacturers will begin to insist that parts and products provided by their subcontractors be produced via ISO 9000 conforming quality systems.

The Basis of ISO 9000

The ISO 9000 series of standards for quality systems is built on the premise that *if the production and management system is right, the product or service that it produces will also be right.*

The ISO 9000 standards are written generically—they are not written for any particular business. As a result, the standards provide models, not specifications. They describe *what*, at a minimum, must be accomplished—they do not specify *how* things must be done. The standards leave it to the manufacturer/developer to decide how to do things effectively. Although it is the purpose of ISO 9000 to standardize quality systems to meet its requirements, it is not its intention to make all quality systems the same.

The ISO 9000 series of standards basically requires that you say what you do, do what you say, and demonstrate what you've done. In other words, document and record! You should have written procedures that define how each significant activity in your development process is conducted. These procedures should be in sufficient detail to allow you to continue making your product or providing your service at the same level of quality if all of your personnel were replaced.

The standards place strong emphasis on *control, auditability, verification/validation*, and *process improvement*. These are some of the key characteristics that your processes and procedures must possess to conform to ISO 9000.

Control means that you are able, at all times during the process, to identify each item, who owns each item, its status, where it is, its currently valid version, proposed changes, etc. To be controlled or under control means *not* being out of control! Auditability means that you can always show objective evidence of what has been done, how it was done, the current status of the project and product, and what is planned to be done, and you can demonstrate the effectiveness of your quality system.

The standards rely heavily on step-by-step verification of results eventually leading to validation that the delivered product meets its specified requirements and is what it was intended to be. This is intended to discourage reliance on mass inspection of a final product as a means of assuring a quality product. ISO 9000 also requires that you continually improve your process, thereby, leading to ever-improving products and services.

ISO 9000 Registration

The approach being used with ISO 9000 is for third-party assessment, via audit, to confirm that your quality system conforms with the ISO 9000 standards. This process is known as ISO 9000 registration (known as ISO 9000 certification in Europe). Registering your quality system assures you and tells the world that you have a quality system that conforms with the ISO 9000 standards. It reduces or eliminates the need for individual purchasers or customers trying to do their own assessment of the supplier's quality system.

Fulfilling the requirements of ISO 9000 can be costly in time, effort, and money. The cost primarily depends on how close your current quality system conforms to the ISO 9000 standards. Making significant changes to an existing process can be disruptive. Simple documentation of existing procedures of a currently conforming process can be expensive and time-consuming.

How long will it take to obtain ISO 9000 registration? The answer depends on a number of factors, including the resources you are willing to commit to the effort. To register you must be able to demonstrate a working (i.e., mature) ISO 9000 conforming quality system. To do this, you will need a minimum of three months of actually using your quality system and keeping records prior to a third-party registration audit. Unless your quality system is already close to conforming and ready for registration assessment, you can expect ISO 9000 registration to take at least one year to 18 months. If you are starting with no system or a poorly documented system, it could take 18 to 24 months or more.

ISO 9000 registration is granted to an organization (*not* to products!) when a third-party accredited registrar assesses the organization's quality system and finds that it does, in fact, conform to the ISO 9000 standards. The assessment is done via an audit (called a registration audit), which involves bringing a registrar's auditors on-site to audit your quality system. The audit consists mostly of interviews with people involved in owning and using the quality system to determine what the processes and procedures are, if they are documented, and if, in fact, they are actually being used. The auditors will report any observed nonconformities and then, based on their judgment as to the seriousness of any nonconformities, recommend for or against registration to their board of directors. Registration is often granted even when nonconformities are present if the totality of nonconformities does not constitute a breakdown in the system or show serious disregard for the standards. Continuing third-party audits (called surveillance audits) are required every six months, and reregistration is required every three years. This could vary slightly depending on your registrar.

The company or organization desiring ISO 9000 registration chooses the registrar it wants to use, applies for registration, and pays the registrar's fee. When registration is granted, the company receives a registration certificate specifying the scope of the audit (i.e., which of the company's quality systems was registered and to which level of the standards) and the name of the registrar that performed the audit and granted the registration.

Other Benefits of ISO 9000

ISO 9000 is not only for those interested in ISO 9000 registration. The ISO 9000 series of standards is a catalyst to quality. It raises quality awareness. If your company already has an established quality management system but is not necessarily interested in registration, it can still be worthwhile for your company to compare its quality system against the ISO 9000 standards. Remember, ISO 9000 provides a basic set of requirements for a quality system. Use ISO 9000 as a measuring stick against which to compare your company's quality system. If your quality system does not stack up well against the standard, you should give serious thought to improving your quality system.

If your company does not have a formal quality management system, the ISO 9000 standards provide an excellent starting point and framework against which to define your quality system. And, for those shooting for quality leadership, the ISO 9000 standards provide an excellent first step toward that objective.

With a focus on software development, the remainder of this book delves into more details about the ISO 9000 standards, suggests key characteristics of ISO 9000 conforming quality systems, makes some practical suggestions on what is required to satisfy the standards, and describes how to proceed with pursuing ISO 9000 registration.

What Is ISO 9000?

ISO 9000 is a series of international standards for quality management systems developed by the International Organization for Standardization. The ISO 9000 series of standards applies to quality management systems for development and manufacturing and defines the minimum process requirements that must be met to assure customers a good product. For the first time, there is broad agreement about the basic requirements for quality systems.

The ISO 9000 series of standards contains three individual but related standards that apply to quality management and quality assurance. The standards are generic and not specific to any particular product. Although originally written for the manufacturing industry, they are also being applied to the service industry. Likewise, the standards are being applied to software development, but the application of some of the elements of the standards to the software development process is challenging and sometimes perplexing. The standards often are found to be unintelligible to software developers; they can read them and still not know what to do next.

The standards provide a framework for quality systems, but they do not specify the particulars for implementation. The standards are *not* specifications—they state *what* has to be done; they do not state *how* things have to be done. The standards leave it to the development organization to design and implement its own effective development process within the framework of the standards. The standards will compel you to build quality into your product or service, thereby reducing the need for costly after-the-fact inspections, warranty costs, rework, and other inefficiencies.

Quality management systems have become a necessity for the survival of any company and the ISO 9000 standards provide a framework upon which to build that system. Regardless of whether registration to ISO 9000 is your objective, ISO 9000 conformance can be very important to you.

The ISO 9000 series of standards consists of five sections: ISO 9000, ISO 9001, ISO 9002, ISO 9003, and ISO 9004. The standards are specified by ISO 9001, ISO 9002, and ISO 9003 with ISO 9000 providing guidance related to which standard to use and ISO 9004 providing amplification and guidance for implementing the standards. The ISO 9000 series of standards provides for three levels of quality assurance.

ISO 9000 Quality systems—Quality management and quality assurance standards—guidelines for selection and use. ISO 9000 provides guidance for selecting which of the standards, ISO 9001, ISO 9002, or ISO 9003, applies to your company's quality system. *ISO 9000-3: Guidelines for the application of ISO 9001 to the development, supply and maintenance of software* provides guidance for the application of the standards to the software development process (see Appendix H).

ISO 9001 Quality systems—Model for quality assurance in design/development, production, installation, and servicing.   ISO 9001 covers processes encompassing product design/development, production, installation, and servicing. If your software development organization designs the product that it develops, then, more than likely, ISO 9001 applies to you.

ISO 9002 Quality systems—Model for quality assurance in production and installation.   ISO 9002 covers processes encompassing production and installation. If your software development organization implements products from a design that is provided to it, then, more than likely, ISO 9002 applies to you.

ISO 9003 Quality systems—Model for quality assurance in final inspection and test.   ISO 9003 covers processes encompassing final inspection and test. If your organization is a test organization, then, more than likely, ISO 9003 applies to you.

ISO 9004 Quality systems—Quality management and quality systems elements—guidelines.   ISO 9004 provides guidance on the interpretation of ISO 9001, ISO 9002, and ISO 9003.

Because ISO 9001 covers more aspects of development, more elements of the standard apply to ISO 9001 than to ISO 9002 and ISO 9003. Figure 1.4 shows how the three standards relate to each other and to the development process. It should also be noted that some of the elements that apply to ISO 9002 and ISO 9003 are less stringent than the corresponding elements in ISO 9001. No sequence is implied by the standard's numbers—they are not a stepwise progression. You do not go to ISO 9001 first, then to ISO 9002, and finally to ISO 9003. However, ISO 9001 is a superset of ISO 9002, which in turn is a superset of ISO 9003. Here we will focus on ISO 9001 because it is the most comprehensive of the standards and is most frequently applied to software development. In most cases, the concepts in this book are equally valid for ISO 9002 and ISO 9003, which are subsets of ISO 9001.

**Manufacturing
Process**

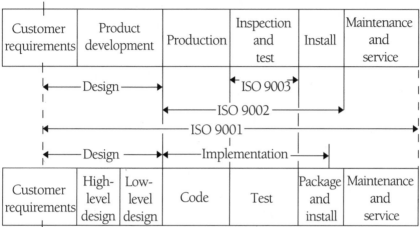

**Software
Development
Process**

Figure 1.4. Relationship of ISO 9000 standards to the development
process.

Why Be Interested in ISO 9000?

There are numerous benefits to having ISO 9000 conforming quality systems.
The ISO 9000 series of standards provides an excellent basis against which to
measure your quality system. Even if ISO 9000 registration is not your objective, and even if ISO 9000 conformance were not needed to be competitive in
Europe and worldwide, you should be judging your quality system against
ISO 9000 to ensure you can consistently provide quality products and service. First and foremost, it is a matter of quality. If quality products are your
objective, then ISO 9000 should be of interest to you.

Other benefits of ISO 9000 include:

A foundation for quality products.   An ISO 9000 conforming quality system will ensure that your development process has a level of control, discipline, and repeatability necessary for you to build quality into your products

consistently, thereby reducing or eliminating the need for more costly after-the-fact inspections, warranty costs, and rework. Continual improvement in your quality system, as required by the standard, will lead to continual improvement in the quality of your product. *You need a quality system to provide quality products!*

Increased productivity and reduced cost.   Many companies that have already implemented ISO 9000 conforming quality systems are finding increases in development productivity. At first this may seem odd, but it actually makes sense. Doing the job right the first time under controlled, repeatable processes reduces the amount of rework and corrective actions required for products produced by less controlled processes. It reduces the amount of wasted time, energy, and money. It also reduces the amount of misunderstanding between and among various developers, which slows down the process. Companies with quality systems that conform with ISO 9000 standards will find productivity improvements over their pre-ISO 9000 quality systems and over their non-ISO 9000 competitors. ISO 9000 results in a better run operation. If your quality system is occasionally inspected or audited by your customer, registering your quality system to ISO 9000 can eliminate the need for these inspections. This is a direct cost savings to both you and your customer.

Consistency.   An ISO 9000 conforming quality system will ensure your customers that not only are you likely to produce the quality product you set out to produce, but you will continue to produce better and better products.

Improved competitiveness.   Conforming to the ISO 9000 quality assurance standards is becoming essential to succeed in an increasingly competitive global marketplace. Having an ISO 9000 conforming quality system will provide a competitive advantage over non-ISO 9000 competitors. ISO 9000 will indicate to potential customers that your products are more likely to meet claims made about them than products with similar claims made by non-ISO 9000 competitors. Further, ISO 9000 is almost a requirement to market and be competitive in the EC, where ISO 9000 is quickly becoming critical to doing business. European companies often favor suppliers that have achieved or are in the process of achieving ISO 9000 registration. Smart U.S. companies can take advantage of this situation by having their quality systems registered to ISO 9000.

Customers demand it.   More and more, as ISO 9000 becomes more prevalent, customers are demanding it. In fact, the ISO 9000 movement is being driven by purchasers who are demanding more assurance that the products they purchase do what manufacturers claim. In contractual situations, increasing numbers of purchasers are specifying that ISO 9000 conforming quality systems be mandatory for the products they purchase. In particular, this often applies to large government bid situations in which the purchaser must protect its investment. Some manufacturers are taking the approach that they are unwilling to lose business because they *lack* ISO 9000 certification.

Keeping up with the Joneses.   If you have an ISO 9000 conforming quality system, you have a competitive edge over your competitors that don't. Conversely if your competitors have ISO 9000 conforming quality systems and you don't, you are at a competitive disadvantage. It's your choice. You may need ISO 9000 just to keep up with your competition.

Corporate image.   ISO 9000 is being used to differentiate between quality companies and the rest of the world. This may be a valid distinction. ISO 9000 conformance requires a demonstrated continuing commitment to quality. A company with ISO 9000 conforming quality systems is a company that is committed to quality!

ISO 9000 is a series of international standards dealing with the development process that are gaining worldwide acceptance and becoming the accepted basis of quality system requirements for product conformity assessment in the global marketplace. The standards are an important ingredient to being competitive worldwide. The ISO 9000 standards are based on the premise that if your production process is right, the product or service produced will be right. The standards specify a minimum set of required activities that must be done during production of a product. Processes that meet the requirements of ISO 9000 standards exhibit control, discipline, and auditability. These characteristics, in turn, instill customer confidence in the supplier's capability to provide products that meet customer expectations.

Although originally written for the manufacturing environment, the ISO 9000 standards are being applied to other industries, including software development. Because of the nature of software and differences between the manufacturing process and the software development process, application of the standards to software development is not straightforward. This book focuses on the application of ISO 9000 standards to the software development environment.

Notes

1. Quality management system, quality assurance system, and quality system are used interchangeably throughout this book and are defined to be the "system" of people, processes, procedures, tools, disciplines, practices, etc., that are directly involved in producing a product or providing a service. They are systems for managing quality, i.e., building quality into a product (see Appendix A).

2

The Standards

Overview

This book focuses on ISO 9001, which is the most comprehensive of the three standards. ISO 9002 and ISO 9003 are subsets of ISO 9001. Although the standards are specified in terms of standards elements (ISO 9001 has 20 elements, ISO 9002 has 18 elements, and ISO 9003 has 12 elements, see Figure 2.1), the underlying essence of the ISO 9000 series of standards is that your quality system must be

- Documented
- Controlled
- Auditable
- Monitored
- Improved
- Effective

In addition, management must be committed and employees must be involved. Simply stated, if your processes are documented, controlled, auditable, effective for you, continually monitored, and improved, *and* your management is committed and your employees are involved, you are well on your way to meeting the requirements of the ISO 9000 series of standards.

The ISO 9000 series of standards specifies *what*, at a minimum, must be done; it does not specify *how* things must be done—that is left up to the development organization to determine. However—and this is key—your process must be *effective* and you must be able to demonstrate that it is effective.

Standard Elements of ISO 9000

The ISO 9000 series of standards specifies 20 standards elements. ISO 9001 requires conformance to all 20 of the elements, while ISO 9002 and ISO 9003 require conformance to 18 and 12 elements, respectively (see Table 2.1). Some of the elements required by ISO 9002 and ISO 9003 are also less stringent than the corresponding elements in ISO 9001.

13

Table 2.1.   Standards levels versus standards elements.

ISO 9000 Standards Elements	ISO 9001	ISO 9002	ISO 9003
4.1 Management responsibility	X	X	X
4.2 Quality system	X	X	X
4.3 Contract review	X	X	
4.4 Design control	X		
4.5 Document control	X	X	X
4.6 Purchasing	X	X	
4.7 Purchaser-supplied product	X	X	
4.8 Product identification and traceability	X	X	X
4.9 Process control	X	X	
4.10 Inspection and testing	X	X	X
4.11 Inspection, measuring, and test equipment	X	X	X
4.12 Inspection and test status	X	X	X
4.13 Control of nonconforming product	X	X	X
4.14 Corrective action	X	X	
4.15 Handling, storage, packaging, and delivery	X	X	X
4.16 Quality records	X	X	X
4.17 Internal quality audits	X	X	
4.18 Training	X	X	X
4.19 Servicing	X		
4.20 Statistical techniques	X	X	X
	20	18	12

Because the ISO 9000 series of standards was written for the two-party contractual situation, the terms "supplier" and "purchaser" are used throughout the standards. Supplier always refers to the party whose quality system is under review. The purchaser is the customer and the subcontractor is always the party from which the supplier is buying a part or product.

Following is a list of the 20 elements of ISO 9001. For consistency and ease of reference, they are numbered here and throughout the book the same as they are numbered in the ISO 9000 standards documents.

4.1 Management responsibility
4.2 Quality system
4.3 Contract review
4.4 Design control
4.5 Document control
4.6 Purchasing
4.7 Purchaser-supplied product
4.8 Product identification and traceability
4.9 Process control
4.10 Inspection and testing
4.11 Inspection, measuring, and test equipment
4.12 Inspection and test status
4.13 Control of nonconforming product
4.14 Corrective action
4.15 Handling, storage, packaging, and delivery
4.16 Quality records
4.17 Internal quality audits
4.18 Training
4.19 Servicing
4.20 Statistical techniques

Following is a summary of the elements of ISO 9001. For each element, the first paragraph provides a more literal statement of the requirement, while the second paragraph paraphrases the requirement in terminology more understandable in the software development environment. Relevant comments on each element are also included.

4.1 Management responsibility. You must have a quality policy for your organization that is understood and implemented throughout the organization. The quality policy must clearly define responsibilities, authority, and interrelations of those with direct influence on quality. Required verification activities need to be identified and you must have resources (qualified personnel and money) for that activity. A manager who is responsible for your quality system must be assigned; and management must periodically review your quality system to ensure its continuing effectiveness.

Management is responsible for establishing a quality policy and committing to it. A quality representative must be assigned who has authority to make decisions and get things done relative to the quality system. Also, management is responsible for communicating and making sure people are aware of and understand the quality policy. This requirement pertains to all manufacturing environments, including software development, and emphasizes that quality must be a major objective and that quality goals and objectives must come from the top of the organization. Assigning a management representative with authority to get things done is one way management can demonstrate its commitment to quality.

4.2 Quality system.   You must establish, document, implement, and maintain a quality system that conforms with the ISO 9000 series of standards. You must have a *documented* quality system in place that enables you to deliver a quality product. Although not specifically required by the ISO 9000 standards, a quality manual is often used to document the quality system and helps to satisfy this element of the standards. (Revisions to the ISO 9000 standards, scheduled for late 1993, may make a quality manual mandatory.)

The standard states, "The supplier shall establish and maintain a documented quality system as a means of ensuring that product conforms to specified requirements." If a quality product is defined as one that conforms to specified requirements, this element of the standard ensures the delivery of a quality product. Recognize, though, that many of the characteristics of quality are often *not* included in the written, specified requirements.

4.3 Contract review.   You must have procedures for ensuring that what is expected from you is adequately defined and documented and that you have the capability to satisfy the requirements.

There must be procedures to ensure that what is expected is clear and understood so the expected results can be produced.

This element was clearly written for the supplier/purchaser contractual situation and specifies that the requirements of the contract must be understood and agreed by both parties. Contract requirements include both what the supplier will do (i.e., supplier requirements) and what the supplier will provide or produce (i.e., product requirements). It also includes what the purchaser will do and provide, if anything.

When a product is produced on speculation rather than as a result of a contractual agreement, the meaning of this element is less clear. Here the product requirements are often provided by an organization other than the product developer or, sometimes, even by the developer. This element requires that the developer have procedures to ensure that the requirements—both supplier requirements and product requirements—are understood and agreed upon regardless of who provides them.

The standard does not require that the product requirements be what the market needs, what will be successful, etc. The standard does not require that the product be successful. In the contractual situation, the purchaser supposedly knows what he or she wants or needs. In the speculative situation, the provider of the product requirements supposedly knows what he or she wants to build. The standard does not require that the supplier produce successful products. Therefore, satisfying ISO 9000 standards does not guarantee success.

4.4 Design control. You must have procedures for controlling and verifying the design output to ensure that specified requirements will be met. You are also required to plan each stage of development activity, plan and carry out design reviews, assign the design verification to qualified personnel, and have defined procedures for controlling design changes.

There must be documented procedures for the design and changes to the design to ensure that the product design meets the specified product requirements. Plans for the design and development activities are also required.

The emphasis of this element is on verifying that the results of the design activity meets the expectation for the product. The expectations for the product are stated as the product requirements, which must be understood by the supplier (see *4.3 Contract review*). The intent is to avoid proceeding into production with a design that will not produce the expected results. In the world of software, all too often the final product turns out to be something other than what was originally intended. Conformance to ISO 9000 standards, and to this element and *4.3 Contract review*, in particular, should help prevent this.

4.5 Document control. You must have defined procedures to control all documents, including review, approval, and change, and to ensure that the right level of information is available to the right people at the right time. You are also required to maintain a master list of current documents.

For documented items, there must be procedures for review, approval, change, issue, and ensuring they get to the right people at the right time and that obsolete documents are not in active use. While this element may seem overly rigorous, it is intended to ensure that everyone on the project is working from the same level of document, such as specifications. Conformance to this element should ensure this. It is, therefore, important to have procedures that ensure obsolete information is not being used.

Identify all documentation that needs to be controlled. This should include all internal documentation that will affect the product and the quality of the product as well as any product documentation that will be produced. The names of controlled documents should be kept on a master list along with the name of each document owner, date of last update, review status, approvals, etc., as a way of maintaining and demonstrating control.

4.6 Purchasing.   You must ensure that parts, obtained from elsewhere, used in the product or in the production of the product, meet their specified requirements. You are also required to select your subcontractors based on their ability to meet the requirements of the subcontract.

If parts are obtained from outside the organization, ensure that they work as expected before using them. Subcontractors must also be selected based on their ability to produce what is expected from them. The ISO 9000 series of standards clearly places the responsibility for assuring the quality of the total product on the organization that delivers the final product (i.e., the supplier). The supplier is responsible for the quality of all the parts of the product regardless of where or by whom they were produced. The intent of this element is to ensure that the supplier validates all parts that go into the product when the parts are received (i.e., before they are actually used in the product). It would be helpful if the provider of all included parts (i.e., subcontractors) were also ISO 9000 registered.

Keeping a list of acceptable subcontractors along with procedures and criteria for adding subcontractors to and removing subcontractors from the list are necessary to satisfy this requirement. In evaluating subcontractors, consideration can be given to factors in addition to the subcontractor's previous history of performance and ability to produce the expected results, such as financial security, market position, and customer satisfaction policies.

4.7 Purchaser-supplied products.   You must have procedures for verification, safe storage, and maintenance of products, or parts, provided by the customer to be included in the product.

If you are modifying a product for a customer, when you receive the product you must verify that it is as expected. From that point on, you must treat it as if it were your own. This standards element requires the developer to verify that what was provided by the purchaser is what was expected. If you do not know what you expected or cannot verify it, how can you possibly plan what has to be done?

4.8 Product identification and traceability. Where appropriate, you must have procedures for identifying and tracing the product during all stages of production, delivery, and installation.

At any time during production, you need to know what parts compose your product, where they are, the status of each, and the specification each supports.

This element was written for the manufacturing environment, more specifically for the assembly line environment, where a number of parts are often brought together to produce some quantity of an end product. For example, an automotive assembly line uses a large number of parts to produce a large number of automobiles. In this environment, it is necessary to track which parts went into which end products. This is especially important if it is later discovered that certain parts are defective, in which case, you must be able to identify the exact products which contain the defective parts.

The software development environment differs in that it brings together a number of parts, which themselves often are being developed, to produce a single final product, which may later be replicated. This element of the standards applies to both environments.

The best application of this requirement to the software development environment is that you (1) control the constituent parts of your product during development and (2) know the exact content of products that have been delivered to customers so that you can provide the right service to customers who have different versions of the product.

Although this element is required only if the developer feels it is needed or if it is specified by contract, it is recommended that you satisfy this element of the standard for your software development environment. Satisfying this element also helps to satisfy element *4.9 Process control*.

4.9 Process control. It is required that you carry out production under controlled conditions, including, monitoring progress, approval of processes and equipment, etc.

Your development process must be controlled. This element is a key element in the standards, yet little is said about it. Several things need to be said.

First, what is production? This element was written for the manufacturing environment where production refers to the assembly process. In software development, there is no clear analogy to production in the manufacturing sense. Conceivably it could be the process of building or integrating the various parts into the final product or it could be the process of putting the final product onto some media for delivery or it could be a number of other things.

The best analogy to production in the software development environment is the set of activities that follows design completion and ends with product delivery (often referred to as implementation).

Second, what does it mean to be controlled? Being controlled means *not being out of control*. It is often relatively easy to recognize when something is out of control while it is more difficult to show that something is controlled. For software production, control means that for all items relating to the product being developed or to the production of the product, there is an owner, with ultimate authority to make decisions and procedures for appropriate review, approval, change, distribution, recall, etc.

Third, this element of the standard makes provision for special processes, that is, processes for which results cannot be fully verified by inspection and testing and process deficiencies may become apparent after the product is in use. Until there are ways to produce zero-defect code, it seems that software falls into the category of special processes. Special processes carry the additional requirement that they be continuously monitored and/or that documented procedures be strictly followed to ensure the specified input requirements are being met. This can be achieved through step-by-step verification as the product progresses through various stages of the development process and with adherence to the documented processes and procedures.

4.10 Inspection and testing.   You must have procedures for all levels of inspection and testing that *you* have identified as being required. You are also required to maintain records of test activity.

You must identify the testing that is required, document the required testing, carry it out, and then show that the required testing was successfully completed. The obvious part of this element is that you are responsible for carrying out the required testing and for being able to show that you did it. Less obvious is that you are also responsible for identifying what testing must be done and for being able to show why that testing is required.

4.11 Inspection, measuring, and test equipment.   You must control, calibrate, and maintain inspection, measuring, and test equipment.

This element of the standard was written for the manufacturing environment, and its analogy in the software development environment is not obvious. The standard mentions calibration and safe storage of test equipment. In the software development environment, calibration can be interpreted to mean demonstrating that tools for testing, verification, validation, measurement, etc., can serve their intended purpose. You must be able to show that your test tools serve their intended purpose and that they are under control. Test tools must be validated and controlled. Therefore, any software test tools must be validated prior to their use and must be controlled similar to the way the software parts that make up the product are controlled.

4.12 Inspection and test status.   You must be able to identify the test status of the product throughout the process.

During test stages, you must be able to show the test status of the various parts/product that are being developed. Part of having your process under control is knowing and being able to show the status of all parts at all times. This element focuses on requirements for test status.

4.13 Control of nonconforming products.   You must have procedures for controlling a product that does not conform to its specified requirements.

You must have procedures in place for what to do with a product when defects are discovered. Defects in software products can be discovered both before and after the product is delivered to customers. Procedures for dealing with the product for which defects are discovered after it has been delivered are typically addressed in response to element *4.19 Servicing*.

When defects are discovered in a product before the product is delivered, the most common practice in software development is to rework the product until it passes all required testing. This element of the standards requires that you have procedures for this. If you decide to deliver a product with known defects or nonconformities, you should inform your customers of this.

4.14 Corrective action.   You must have procedures for investigating the causes for nonconforming products and ensuring corrective actions to prevent recurrences.

You must have procedures for identifying things that went wrong or could be improved in your development process or quality system. Some people interpret this element as pertaining to corrective action to nonconforming products (i.e., how you fix "bugs"). The element, however, addresses corrective action to the *process* used to develop the product, not to the product. Procedures for corrective action to products that do not meet the specified requirements (i.e., have defects) are addressed in *4.13 Control of nonconforming product* and *4.19 Servicing*. This element ensures that you continually monitor, assess, and improve your development process and quality system. It is likely that you will need to keep records of product defects/problems and customer complaints to satisfy this requirement. You will likely also need to gather and keep records of development process metrics.

4.15 Handling, storage, packaging, and delivery. You must have procedures for handling, storing, packaging, and delivery of the product.

You must have a good system for storing and controlling the various parts that will compose your product during product development and through product delivery.

This element appears to have been written for the manufacturing environment where the product often requires special handling and storage to protect it until it is in the customer's hands. In software terminology, this requires that all parts and associated work items be stored in a safe, secure, and controlled place. It also requires that you have procedures to ensure that you ship the product—and only the product—you intend to ship and that your customer receives what you send. This verification must be done prior to customer installation. You must also state and agree beforehand who performs the verification, you or the customer.

If you use a central computer facility during the development process, environmental control of this computer facility is covered by this element.

4.16 Quality records.   You must identify and keep records to demonstrate achievement of product quality and effective operation of your quality system.

You must identify and keep whatever records are needed to demonstrate and improve the effectiveness of your quality system. The standard does not specify what records must be kept; it requires that you identify what records need to be kept. You should keep whatever records will enable you to demonstrate the effectiveness of your quality system. This usually involves identifying and keeping both product and process metrics.

4.17 Internal quality audits.   You must plan and carry out internal quality audits, by qualified individuals, to verify you are doing what you say you are doing and to determine the effectiveness of your quality system.

You must conduct periodic internal audits of your quality system. Internal audits must be conducted by qualified personnel, and it is up to you to show evidence that those who do the audits are properly trained and qualified. The audits must be carried out according to documented procedures.

Internal audits are used to determine that your quality system is being used and that it is effective. These audits differ from ISO 9000 surveillance audits in that they must determine how effective your quality system is, which is not part of an ISO 9000 audit. For this purpose, internal audits may require pulling together quantified information.

4.18 Training.   You must identify the training needs of your people, provide the required training, and keep records of the training.

To satisfy this requirement, you must know (1) what kind and how much training each of your involved employees has, and (2) what skills are required to perform each task in your process. This information must be recorded for you to demonstrate that you are providing the training your people need.

4.19 Servicing. You must have procedures for servicing your product when servicing of the product is specified in the contract.

If you provide service for your product and it is specified in the contract, you must have procedures for providing that service. This section requires documented procedures for correcting defects and nonconformities found in the product after it has been shipped. It addresses your service process if servicing of your product is provided by your organization and is specified in the contract.

4.20 Statistical techniques.   You must show that any statistical techniques that you use are correct.

You must be able to show that any metrics or measurements you use during development of the product and/or to determine the quality of your product and the effectiveness of your quality system are correct and accurate. In addition to validating the metrics, measurements, etc., you must be able to show validity of any predictive algorithms used (e.g., for predicting number of remaining problems) and methods used to collect data.

These are the 20 elements of ISO 9001 that specify the standards requirements. Some of the requirements can be applied directly to the software development environment with little or no interpretation. Others, however, require considerable interpretation to apply properly to software development. The following chapters provide an interpretation and suggest implementations that will satisfy the requirements of the standards for software development.

3

ISO 9000 Registration

Registration—What Is It and Why Do It?

Corporations around the world are building quality systems based on the ISO 9000 series of standards. At the same time, increasing numbers of customers expect the companies with which they do business to be registered to ISO 9000. Registration involves having an accredited, independent third party (i.e., registrar) conduct an on-site audit of a company's operation against the requirements of the ISO 9000 series of standards. Upon successful completion of a registration audit, the company will receive a registration certificate confirming that its quality system is in conformance with one of the ISO 9000 series of standards. The certificate indicates to which of ISO 9000 standards (i.e., ISO 9001, ISO 9002, or ISO 9003) the company's quality system conforms and identifies the registrar that performed the audit. ISO 9000 registration is awarded to the company or organization, indicating that its quality system conforms with one of the ISO 9000 standards. It is a certificate of which a company can be proud, and it can be used to the company's benefit (e.g., in advertising). It is a stamp of approval on the company's quality system—it is *not* a stamp of approval on a product.

Registration is neither required nor mentioned by the ISO 9000 series of standards. Registration is part of the approach being used with ISO 9000 and is increasingly being demanded by customers.

ISO 9000 registration does not end with the issuance of the certificate of registration. ISO 9000 conformance must be maintained. The ISO 9000 series of standards requires that you perform regular internal audits and that you continually improve your quality system. Registration also requires regular, periodic surveillance audits (known by registrars as continuing audits) that are conducted by the original registrar at the registrar's discretion—usually two per year, generally depending on how conforming your quality system is. Realistically some quality systems are more conforming than others. For example, some companies may have quality systems that have been in conformance for a long time while other companies may have just established a quality system, which, with some wrinkles, conforms. The latter will need to be more closely monitored until the process is better established.

In addition to having periodic surveillance audits, your company will need to have its quality system reregistered about every three years, again depending on your registrar. Reregistration involves a full registration audit conducted by a registrar, usually the original registrar. You may, however, choose to change registrars.

So, if you have an ISO 9000 conforming quality system, why bother to get registered? There are several good reasons. The certificate tells the world that not only does your quality system conform to the ISO 9000 series of standards, but an accredited third party has confirmed it. And now you have a certificate of registration to prove it. You can flaunt it, advertise it, and use it to your competitive advantage.

The EC is providing the impetus to many companies for ISO 9000 registration. Several EC directives will require manufacturers of regulated products (i.e., those associated with safety, health, and environment) to have a documented and well-implemented quality system, as a minimum. More and more companies are using the ISO 9000 series of standards to document, implement, and demonstrate their quality systems for both regulated and unregulated products. Some people believe that ISO 9000 registration will become essential to being competitive in the European market. ISO 9000 standards and the third-party registration scheme are also gaining acceptance worldwide.

In situations where customers require in-process inspection of a supplier's process to satisfy themselves that they are going to get the product they expect, ISO 9000 registration can supplant the necessity for such second-party audits, thereby avoiding the expense and disruption caused by on-site inspection. ISO 9000 registration provides assurance to customers that their supplier has achieved a basic acceptable level of control and is committed to providing a quality product. Customers, in increasing numbers, are expecting their suppliers to be ISO 9000 registered.

Finally, after having done what has to be done to conform to the ISO 9000 standards, why not go the final step and get registered? The cost of the registration process, although not trivial, is far outweighed by the benefits. Although not required by the standard, registration is the approach being used worldwide by suppliers to let their customers and competitors know that they have a quality system that conforms to the ISO 9000 series of standards. While registration to ISO 9000 may not guarantee that a product will meet all of a customer's expectations or will be of the highest quality, it does improve the customer's confidence of receiving a quality product that meets their expectations.

4

Characteristics of an
ISO 9000 Quality System

There are seven essential characteristics of any ISO 9000 conforming quality system. These are characteristics without which you can have neither a good quality system nor an ISO 9000 conforming quality system. Some of these characteristics are attitudinal and others are basic, underlying attributes of any good quality system. They are as follows:

1. Quality objectives. First and foremost you must have quality objectives. Your company should have a quality policy that states its quality goals and objectives and the strategy it will use to achieve them. What are your company's quality goals? What are you trying to achieve? Your quality policy should answer these kinds of questions. Your quality policy can be brief, and the objectives should be ambitious yet achievable. Don't set quality objectives that are unreasonable and cannot be met.

   Your quality policy needs to be established and supported by your highest levels of management and must be known and understood by all employees (at least those directly involved in development of your product and your product's quality).

2. Commitment, involvement, and attitude. To have an ISO 9000 quality system means that all employees and managers must be committed to the quality objectives and involved in achieving the objectives. Top-level management is responsible for managing quality and must visibly demonstrate its commitment to the quality objectives, and all employees must be involved in putting the quality system into practice. Everybody must follow the documented procedures, follow the same process, and strive for the same objectives.

3. Controlled. Controlled may be the key word in this entire discussion. Every aspect of what is done during the development process must be controlled—or *under control*. If you have a process that fosters control, you have one of the main ingredients of a good quality system and an ISO 9000 conforming quality system. If there are some development

activities or some items being developed that are not under control (i.e., they are out of control), then you will have to bring them under control if you hope to achieve ISO 9000 conformance. You may have to add procedures or redefine processes to add the required levels of control.

What does it mean to be controlled? For every item related to the product or the development of the product (e.g., product requirements, specifications, test plans, schedules, product parts, product documentation, process, and procedures), there needs to be: (1) an owner with the authority to approve changes; (2) procedures for requesting, reviewing, and approving changes (i.e., a change procedure); and (3) review procedures for validating the item.

Changes need to be requested, explained, reviewed with and by other affected parties, discussed, decided upon, documented, and then communicated in a timely manner to all people who need to know. With this level of control on all items that change during development, your process will have the necessary element of control.

4. Effective. Effective may be the second most important word in this discussion because it holds the clue to many of the value judgments made about your development process and your quality system. It is the means by which you measure whether your quality system is really working for you. It is also the way auditors determine if the way you do things is acceptable. You will need to know how effective your process is in order to (a) demonstrate its effectiveness, (b) improve its effectiveness, and (c) determine your cost of quality. ISO 9000 auditors will want to be shown that your quality system is effective for you. They should not be interested in how effective it is, however, only that it is effective.

Your quality system should be designed for your business and around your quality objectives, not for the standard and certainly not just to satisfy the auditors. It should be designed to be effective for you and for what your company does. It should be only as comprehensive as is necessary to meet your quality objectives. Don't include things unnecessarily just to satisfy the standards. Conversely if there are quality considerations that you need, include them in your quality system even though they may not be required by the standards.

The ISO 9000 series of standards requires that you demonstrate the effectiveness of your quality system. The standards do not prescribe any particular process or procedures that must be used. They merely state that your quality system be effective. And you are the best judge of that. But you must be able to demonstrate that it is effective. There are numerous ways to do that, including, showing schedules have been, or are being, met, showing improved productivity, showing improved product quality, showing improved customer satisfaction, showing reduced cost, etc. All of these require proper documentation and record keeping.

The ISO 9000 standards do not prescribe the processes or procedures that you must use. You decide how to do things. The standards require you to show that your quality system is effective.

5. Auditable. The ISO 9000 series of standards requires that systematic internal audits of your quality system be conducted. ISO 9000 registration also requires that continuing audits (i.e., surveillance audits) be conducted periodically by independent third-party auditors. ISO 9000 also requires that your quality system, including procedures, be documented. It emphasizes documentation of what you plan to do and recording of what you've done. All of this constitutes auditability. Your quality system must be auditable. At any point during the development process, you must be able to show where you are, what has been done, and what has yet to be done. You should record only what needs to be recorded. A good rule-of-thumb is to record only what you will need to refer to later.

A quality system, which is auditable, makes it easy for you to satisfactorily answer many of the questions posed during an audit. It also facilitates demonstrating the effectiveness of your quality system.

6. Documented quality system. Your quality system, including your processes and procedures, should be documented to the extent that, if you had to replace all of your employees, you could do it and still continue your business. All procedures that are not readily intuitive to the person performing them or that involve more than one person should be documented. Procedures intuitive to the person performing the procedure do not have to be documented, unless someone else needs to know how it is being done. You should document only what needs to be documented.

7. Continual improvement. The ISO 9000 series of standards requires that your quality system be continually monitored and reviewed for weaknesses and that improvements be identified and implemented.

This is done with the notion in mind that improvements in the process will ultimately lead to improvements in the quality of the product produced by that process.

These are the basic characteristics of a good ISO 9000 conforming quality system. If your quality system possesses all or most of these elements, you have a good basis for an ISO 9000 conforming quality system. If your quality system lacks any of them, you should seriously consider making whatever changes are necessary to bring the missing characteristics into your quality system to improve it. You will also need to make the changes to become ISO 9000 registered.

5

Satisfying the Standards

If your quality system possesses the basic characteristics outlined in the previous chapter, you are well on your way to conforming with the ISO 9000 series of standards. Now the challenge is to manifest these characteristics in actual practice. You need to be able to show that your quality system adequately addresses all the elements of the standards and to demonstrate that what you are doing is efficient and effective for you.

This chapter proposes a number of items that, together with the characteristics outlined in Chapter 4, will satisfy the requirements of the ISO 9000 standards for software development. A description of each item is provided along with an explanation of which elements of the ISO 9000 standards the item addresses and how. This approach is used rather than addressing the standards element-by-element, which can lead to a less understandable quality system and often raises unnecessary questions about how some of the elements really relate to various aspects of the software development process.

The following items will provide a controlled software development process and satisfy the requirements of ISO 9001 for software development.

- Quality policy
- Quality manager
- Quality manual
- Documented processes and procedures
- Project plan
- Build plan
- Test plan
- Service plan
- Quality records
- Training records
- Internal quality system audits
- Library control system

Following are descriptions of each of these items.

Quality Policy. You must have a quality policy in written form.[1] The quality policy states your organization's quality goals and objectives and the strategy

31

for achieving them. The quality policy must be advocated and supported by the highest levels of management. A documented quality policy, signed by your highest level of management, serves this purpose. It must be more than lip service by the person who signs it. ISO 9000 auditors will check for management awareness of and commitment to your quality policy.

Excerpts from existing quality policies are included in Appendix B for your consideration and, if appropriate, your judicious use. Quality policies do not have to be lengthy to be effective.

The ISO 9000 standards require a quality policy (*4.1 Management responsibility*, specifically *4.1.1 Quality policy*). The quality policy described here satisfies this element of the standards.

Quality Manager.   You must assign a management representative, reporting at a high level, to be responsible for your quality system and for assuring ISO 9000 conformance. It is often convenient, as well as sensible, to assign other quality responsibilities to this manager, such as planning internal audits, ownership of the quality system, ownership of quality records, periodic quality system assessments, employee awareness, and management reviews of the quality system. Your quality manager can have other responsibilities in addition to his or her quality responsibilities.

The ISO 9000 standards require that a management representative be assigned with authority and responsibility for ensuring that the requirements of the ISO 9000 standard are implemented and maintained (*4.1 Management responsibility*, specifically *4.1.2.3 Management representative*). It also requires that a comprehensive set of planned and documented internal audits be carried out (*4.17 Internal quality audits*).

The quality manager described here satisfies the requirement for a management representative and contributes to satisfying the requirement for internal quality audits by being responsible for planning and carrying out internal audits and taking timely corrective actions that result from the audits.

Quality Manual.   The ISO 9000 standards require that your quality system be documented. Although not specifically required by the standards, a quality manual, together with documented processes and procedures, serves this purpose and addresses this requirement of the standard. The quality manual provides a blueprint of your quality system while the documented processes and procedures (discussed later in this chapter) provide the details. In addition to being a place to state your quality policy, the quality manual should address each of the ISO 9000 elements. It should state convincingly how your quality system addresses or responds to each of the elements. For elements that do not apply to your organization, you should state why they do not apply.

The quality manual is a key item. It may be the only document a registrar looks at prior to the registration audit. A good quality manual can have a positive influence on the audit team even before it arrives for the audit. It must be thorough and convincing. After reviewing your quality manual, an auditor should be convinced that your quality system adequately addresses each of the ISO 9000 elements. When addressing the elements, do not simply reword each element requirement. Do not take the requirement "supplier shall inspect, test, and identify product as required by the quality plan . . ." and change it by stating that you "inspect, test, and identify product as required by the quality plan." This is not convincing, especially if the entire quality manual is full of these reworded requirements. Provide enough detail to convince the reader that you have seriously addressed the requirement. Time spent on the quality manual is time well spent.

A quality manual can be a multipurpose document. It can also serve as a vehicle for communicating your quality policy, goals, and objectives to all of your employees. If it is written thoughtfully and carefully, it can also be used to advertise your quality initiatives and accomplishments to prospective employees and customers.

A quality manual, per se, is not required by ISO 9000, and the requirement to have a documented quality system can be satisfied in other ways. However, a well-written, well-thought-out quality manual, organized to address each of the ISO 9000 standards elements, makes the auditors' job easier and can make a favorable impression on the auditors. It is unlikely that you can fail an audit for lack of a quality manual, but you must have your quality system documented to satisfy the ISO 9000 standards. Why not develop a quality manual and use it to your best advantage? (Revisions to the ISO 9000 standards, scheduled for late 1993, may make a quality manual mandatory.)

The ISO 9000 standards require that your quality system be documented (*4.2 Quality system*). It also requires that management be responsible for ensuring its quality policy and objectives are understood throughout the organization (*4.1 Management responsibility*, specifically *4.1.1 Quality policy*). A quality manual as described here provides a portion of the documented quality system and also serves as a vehicle for communicating the quality policy and objectives to your organizations and others.

A quality manual is subject to change and, therefore, must be controlled. To be under control, it needs an owner, a change procedure, etc.

An outline for a quality manual is provided for your use in Appendix C.

Documented Processes and Procedures.   A commonly asked question is what procedures need to be documented. A general rule is that you should document all procedures that would be needed to continue your operation if all of your people were replaced. The overall process and the more detailed procedures used throughout the development process must be considered for documenting. They are an important part of your quality system. Specifically, when deciding whether to document or not, consider the following questions.

1. Does the procedure directly or indirectly affect the quality of the product being developed? If so, it probably should be documented.

2. Is there more than one way to do this *and* is more than one person involved? If so, it probably should be documented.

3. Is it intuitive to the person performing the procedure? If so, you may not want to document the procedure. When people possess the skills that allow them to act independently, make decisions, and apply their skills and knowledge toward being creative and productive, documented procedures for these activities could be more of a hindrance than a help.

When in doubt, document!

The required qualifications of the person performing the various procedures should also be listed as part of the procedure description. This will help to satisfy the training requirement (see *Section 4.18 Training*). It will also be helpful if you are asked by an auditor, "How do you know if this person has the necessary skills to perform this task?"

Each documented procedure should specify, at a minimum, the following:

- Expected input and acceptance criteria
- Expected output
- Interrelationship with other procedures
- Qualifications and skills of the person performing the procedure
- Tools, rules, practices, methodologies, and conventions used

Your own people should be involved in defining and documenting your procedures. They are the ones who know or need to know the procedures. They are the ones who will use them. They will benefit the most from this exercise. A consultant can help with this and can provide valuable guidance in establishing procedures that conform to the standards. Appendix D provides a template that can be used as a guide for documenting procedures.

The ISO 9000 standards require that you have a documented quality system (*4.2 Quality system*). The documented processes and procedures provide much of the documentation of the quality system. Documented processes and procedures, together with the quality manual outlined earlier, form the bulk of the documented quality system.

The ISO 9000 standards require that you have procedures for the following: contract review (*4.3 Contract review*); control, verification, and change to the design of the product (*4.4 Design control*, specifically *4.4.1 General* and *4.4.6 Design change*); control of documents (*4.5 Document control*); verification, storage, and maintenance of purchaser-supplied product for inclusion in the final product (*4.7 Purchaser supplied product*); identifying the product during all stages of production, delivery, and installation (*4.8 Product identification and traceability*); ensuring adequate process control (*4.9 Process control*); final inspection and testing (*4.10 Inspection and testing*); ensuring that product that does not conform to specified requirements is prevented from inadvertent use (*4.13 Control of nonconforming product*); investigating the cause of nonconforming product, causal analysis, initiating preventative actions, ensuring corrective actions are taken, implementing and recording changes in procedures (*4.14 Corrective action*); handling, storage, packaging, and delivery of product (*4.15 Handling, storage, packaging, and delivery*); identification, collection, indexing, filing, storage, maintenance, and disposition of quality records (*4.16 Quality records*); audits and follow-up action (*4.17 Internal quality audits*); identifying the training needs and providing for the training of all personnel performing activities affecting quality (*4.18 Training*); performing and verifying that servicing meets the specified requirements (*4.19 Servicing*); and identifying adequate statistical techniques required for verifying the acceptability of process capability and product characteristics (*4.20 Statistical techniques*). Your documented procedures for each of these required areas address these standards elements.

Because processes and procedures are subject to change, their documentation must be controlled. To be under control, the documented processes and procedures need an owner, a change procedure, etc.

Project Plan.   Simply put, plan what has to be done. For software development, this means planning the steps and activities that will be performed in transforming the product requirements into a final product. A project plan should be established at the beginning of a project and updated as the product progresses. The project plan should be a living document showing what is planned to be done. Since plans usually change over the life of a project, so will the project plan. At any point during a project, the project plan will tell you what is planned to be done during the remainder of the project.

The project plan should include the following items.

- Project definition—states the project objectives and what requirements are being addressed.
- Project organization—identifies what project teams, subcontractors, etc., will be used and what pieces of work, activities, or tasks each will perform. It also identifies the sequence of activities, dependencies and the expected output, results, and deliverables from the various activities.
- Project schedule—defines major checkpoints and milestones which are used to review progress (e.g., consumption of resources versus the plan) and technical achievement (e.g., status of expected activity deliverables).
- Project management—describes how the project will be managed, including progress control (i.e., what progress reviews will be held, when, purpose, etc.), organizational responsibilities, resources and skills required, and work assignments.
- Related item—identifies what other plans/documents are related to this project (e.g., quality manual, process and procedure descriptions, test plans, etc.).

The ISO 9000 standards require that you draw up plans for each design and development activity (*4.4 Design control*; specifically *4.4.2 Design and development planning*). The project plan responds to this requirement of the standards. By showing what skills are required and committing the necessary skills and resources to each task, the project plan helps satisfy the standards requirement that you have the capability to meet your contractual obligations (*4.3 Contract review*).

The project plan described here is comparable to the development plan suggested by ISO 9000-3 (see Appendix H). Because a project plan is subject to change, it must be controlled to be under control; it must have an owner and procedures for review, approval, change, distribution, withdrawal, etc.

Build Plan.   A build plan is used to plan, control, and track the various parts that compose a software product, including product documentation. It should specify what parts have to come together to create the total product, in what order, when, and it should specify their interdependencies. The build plan shows the status of each part version at any time during the product life cycle, and it should provide information that ties each version of a part back to a specification (i.e., what function is supported by the part version).

When the product ships, the build plan for the product provides the list of actual part versions that make up the shipped product. The build plan, coupled with a list of customers who have received the product, provides a record of who has what for maintenance and other purposes.

The build plan provides a mechanism for change control, that is, for identifying, tracking, and controlling changes to software items.

A main purpose for the build plan is to bring control and manageability to the software development process. It provides for identification, traceability, and status of the product and its parts as they evolve. As such, the build plan addresses the ISO 9000 standards requirement for identification and traceability (*4.8 Product identification and traceability*). It helps satisfy the standards requirement for controlled conditions (*4.9 Process control*). The build plan makes provision for identifying status of parts and thereby addresses the requirement to identify parts that might be in use before testing is complete (*4.10 Inspection and testing*, specifically *4.10.1.2 Incoming product released for urgent production purposes*).

As a mechanism for keeping status of parts, the build plan is the place where test status of parts and the product is kept. As such, the build plan addresses the ISO 9000 standards requirement to maintain test status throughout production (*4.12 Inspection and test status*).

Finally, the build plan facilitates ensuring that the final delivered product is what it is supposed to be, thereby protecting the quality of the product during delivery as required by the standards (*4.15 Handling, storage, packaging, and delivery*, specifically *4.15.5 Delivery*).

The build plan would constitute part of the configuration management system specified by ISO 9000-3 (see Appendix H). Because the build plan is subject to change, it needs to be controlled. To be under control, the build plan must have an owner and documented procedures for review, approval, change, distribution, withdrawal, etc.

Test Plan. Every project should have a test plan that is established at the beginning of the project and updated as the project progresses. The test plan is a comprehensive document which addresses a number of topics relating to testing and quality activities. The test plan should specify the test strategy that will be used for the project. It serves as a quality plan for the project by specifying the various quality activities that will take place during the project. It identifies the various testing and inspections you have determined to be required for the project. And it provides rationale and justification why the identified testing and inspection is adequate and will serve its intended purpose.

The ISO 9000 series of standards specifies certain times during development when validation/verification[2] is required. However, it is up to the developer to identify how that will be carried out and identify the testing that is required to ensure the product satisfies its stated requirements. The test plan is where you describe the testing that you have determined is needed during the production of a quality product. You must be sure, however, that the planned testing meets, at a minimum, the validation/verification requirements of the ISO 9000 standards.

The ISO 9000 standards require that you identify in-house verification requirements (*4.1 Management responsibility*, specifically *4.1.2.2 Verification resources and personnel*). They also require the following verification: design verification (*4.4 Design control*, specifically *4.4.5 Design verification*), purchased product conformity to specified requirements (*4.6 Purchasing*, specifically *4.6.1 General*), verification of purchaser-supplied product for inclusion into the product (*4.7 Purchaser supplied product*). Several levels of testing are also required (*4.10 Inspection and test*), including verification of incoming products as to conformance to specified requirements (*4.10.1 Receiving inspection and testing*), inspection and testing of product as required (*4.10.2 In-process inspection and testing*), and final inspection and testing (*4.10.3 Final inspection and testing*). All of these activities should be documented in the test plan.

You must be able to demonstrate that test tools are capable of doing what is intended (*4.11 Inspection, measuring, and test equipment*). The test plan provides an explanation of why the various testing is adequate.

The standards require that test status be kept throughout production. Test status can be kept with the test plan; however, in the scenario presented in this book, test status is kept in the build plan.

The test plan described here incorporates much of the quality plan, verification plan, and test plan suggested by ISO 9000-3 (see Appendix H). Because the test plan is subject to change, it needs to be controlled. To be under control, the test plan must have an owner and documented procedures for review, approval, change, distribution, withdrawal, etc.

Service Plan. Every product should have a service plan stating the planned maintenance activities that will be carried out after the product is delivered and who will perform the activities. The service plan should include the following:

- Maintenance procedures
- What types of problems will be corrected under maintenance procedures
- Who will do what
- How errors will be reported
- How maintenance (i.e., fixes) will be provided
- Response criteria
- What information will be collected, reported, analyzed, and fed back to design/development

The ISO 9000 standards require that information, including service reports, be analyzed to eliminate potential causes of nonconforming products (*4.14 Corrective action*). They also require that records be maintained for the purpose of demonstrating the achievement of the required quality and the effectiveness of the quality system (*4.16 Quality records*). The service plan specifies what service-related information will be collected for analysis and evaluation.

The service plan directly addresses the ISO 9000 standards requirement for procedures for performing and verifying servicing the product after delivery (*4.19 Servicing*). The plan explains the usage of service-related information that will be collected. This partially addresses the ISO 9000 standards requirement to identify adequate statistical techniques (*4.20 Statistical techniques*). The service plan is similar to the maintenance plan suggested by ISO 9000-3 (see Appendix H). Because the service plan is subject to change, it needs to be controlled. To be under control, the service plan must have an owner and documented procedures for review, approval, change, distribution, withdrawal, etc.

Quality records.   Unlike the project plan, test plan, and service plan that document what is going to be done, quality records provide a record of what has been done. To a large degree, quality records are kept so you can show that you have done what you said you were going to do. But they serve more of a purpose than just that. Quality records should provide a valuable source of information when you assess the effectiveness of your process and quality system. For example, you can use your quality records to establish the relative product quality from one product to another, the relative product cost, the relative development productivity, etc. You will likely need quality records that pertain to the process, and its effectiveness, and to the product, and its quality.

Quality records can include inspection reports, audit reports, test results, review and inspection reports, error records, change activity, design change review reports, process postmortem reports and action plans, lists of acceptable subcontractors, customer satisfaction, customer complaints, etc. The standards leave it to you to decide what records need to be kept as quality records. At the beginning of each project, you should identify what records need to be kept to enable you to determine the effectiveness of your quality system. A record of any event or data that provides evidence that the process is being followed and/or pertains to the effectiveness of your process should be considered to be a quality record and treated as such. You should record only what needs to be recorded. Record only what you expect to refer to later.

The ISO 9000 standards require that quality records be maintained to demonstrate achievement of required quality and the effective operation of the quality system (*4.16 Quality records*). Numerous other elements of the ISO 9000 standards require that records be kept for later reference and as evidence that the activity was performed. Quality records as described here directly address the ISO 9000 standards requirement for quality records and partially addresses requirements of numerous other elements of the standards.

For your quality records you should specify how they will be kept, where they will be kept, and for how long. Because quality records are subject to change, they need to be controlled. To be under control, quality records must have an owner (often the quality manager) and documented procedures for review, approval, change, distribution, withdrawal, etc.

Training Records.   The ISO 9000 series of standards requires that you be able to show that you assign qualified people to various tasks and that you identify and provide required training to your employees. There are two parts to satisfying this requirement: one is knowing the qualifications and skills required for each task and the other is knowing the qualifications and skills possessed by each of your employees. The first, qualifications required for the task, is achieved by identifying the required skills in each procedures description (see *Documented Processes and Procedures*, pages 34–35). Then you need to maintain training records for each employee indicating what training has been provided, when, and how. Training includes more than formal education courses and classes. It also includes on-the-job training when the training successfully provides the employee with a new skill. All such training must be recorded and kept for each employee.

The ISO 9000 standards specifically require that records of training be maintained (*4.18 Training*) and that records for qualified personnel also be maintained (*4.9 Process control*, specifically *4.9.2 Special process*). Qualified personnel are those people with special certificates, accreditation, licensing, etc. Training records as described here directly address these requirements.

In addition, the standards require that management assign trained personnel for verification activities (*4.1 Management responsibility*, specifically *4.1.2.2 Verification resources and personnel*) and competent personnel for verifying the design (*4.4 Design control*, specifically *4.4.5 Design verification*). They also require that the supplier has the capability to meet contractual requirements (*4.3 Contract review*). These requirements are partially satisfied through the use of training records as described here.

Because training records are subject to change, they need to be controlled. To be under control, training records must have an owner and documented procedures for review, approval, change, distribution, withdrawal, etc. It is a good idea to keep the employee training records with the employee's personnel file. Don't, however, make them part of the file such that they cannot be shown to an auditor.

Internal Quality System Audits. Periodic planned internal audits of your quality system should be conducted by qualified personnel for the purpose of determining the effectiveness of your quality system and ensuring that planned activities and procedures are being followed.

Carefully planned internal audits of your quality system with objectives to (1) determine its effectiveness, (2) identify areas for improvement and corrective action, and (3) verify conformance to documented procedures can serve multiple purposes and, at the same time, help satisfy the ISO 9000 standards. Internal audits are a valuable source of information for your use and are a powerful tool for identifying improvements to your quality system. They provide information for correcting process problems, improving the process, and proving the effectiveness of your quality system. Internal audits must be performed by trained personnel and carried out according to documented procedures. Results of internal audits must be reviewed with management as part of management's responsibility to periodically review the quality system.

The ISO 9000 standards require that you carry out a comprehensive system of planned and documented audits (*4.17 Internal quality audits*). Internal audits of your quality system, as described here, directly address this requirement. The standards also require that your quality system be reviewed at appropriate intervals by management to ensure its suitability and effectiveness (*4.1 Management responsibility*, specifically *4.1.3 Management review*). Results of internal quality system audits provide valuable input to these reviews.

The standards require that you investigate deficiencies in your quality system to prevent recurrences of nonconforming products (*4.14 Corrective action*). Data gathered during internal quality system audits provide valuable information for this activity.

Library Control System.   Only the smallest projects can be managed effectively without some kind of on-line, automated library control system for storing, maintaining, and controlling various quality system items as well as the parts being developed. The ISO 9000 series of standards requires proper and safe storage of the parts being developed. A good library control system provides a safe and secure place to store the parts being developed (i.e., code and product documentation[3]), as well as the necessary levels of control of changes. It also provides for security, backup and recovery, archiving, error recovery, etc. The library control system should also be used to store and control project and quality system documentation including, documented processes and procedures.

Items best stored and maintained in an on-line library control system include code, product documentation, quality manual, quality policy, process and procedure descriptions, quality records (including subcontractor list, project plan, test plan, build plan, service plan, test tools, test cases, and test scenarios) product requirements, product specifications (design output), master document list, distribution lists, library system description, and training records.

A library control system is a good idea no matter what the size of the project. It helps satisfy several of the ISO 9000 standards requirements and provides control and convenient access to parts, documents, and records. For instance, documents (*4.5 Document control*), parts under development (*4.9 Process control*), and test tools (*4.11 Inspection, measuring, and test equipment*) must be controlled. A library control system provides facilities to control documents and parts under development. The standards require that product handling is safe, storage is secure, and packaging is controlled (*4.15 Handling, storage, packaging, and delivery*). A library control system facilitates storage and control of the product being developed.

A library control system would constitute part of the configuration management system specified in ISO 9000-3 (see Appendix H).

This chapter suggests a number of items for each software development project. These items, together with the quality system characteristics outlined in Chapter 4, go a long way toward satisfying the requirements of ISO 9001 for software development. Figure 5.1 presents a summary of the characteristics (explained in Chapter 4) and the items (explained in this chapter) that are essential to satisfying the requirements of the ISO 9000 standards for software development. Figure 5.2 shows the essential characteristics and items and indicates which of the 20 ISO 9000 elements are addressed by each.

These items, along with the guidance offered by ISO 9000-3 (see Appendix H), should be considered when designing and assessing your quality system for conformance to ISO 9001 in the software development environment. The items do not have to be implemented as suggested, and these items, of course, can be combined or named differently. The contents of the items outlined in this chapter and in ISO 9000-3, however, need to be present in your quality system for it to conform to ISO 9001 for software

Characteristics (Chapter 4)	Items (Chapter 5)
Quality objectives	Quality policy
Commitment, involvement, and attitude	Quality manager
	Quality manual
Controlled	Documented processes and
Effective	procedures
Auditable	Project plan
Documented quality system	Build plan
Continual improvement	Test plan
	Service plan
	Quality records
	Training records
	Internal quality system audits
	Library control system

Figure 5.1. Essentials to ISO 9000 conformance for software development.

ISO 9000 Standards Elements

Essentials to conformance	1	2	3	4	5	6	7	8	9	10	11	12	13	14	15	16	17	18	19	20
Quality objectives	X	X																		
Commitment, involvement, and attitude	X												X							
Controlled	X			X	X				X	X	X		X			X	X			
Effective	X	X															X			
Auditable	X			X	X		X				X	X	X		X		X	X		
Documented quality system	X	X																		
Continual improvement	X												X							
Quality policy	X																			
Quality manager	X																	X[a]		
Quality manual	X	X																		
Documented procedures and processes	X	X	X	X		X	X	X	X			X	X	X	X	X	X	X	X	X
Project plan			X[b]	X																
Build plan							X	X	X[c]		X		X							
Test plan	X				X		X	X			X	X[d]								
Service plan													X		X			X	X	
Quality records	X	X	X	X		X	X	X	X	X	X	X		X		X	X	X		X
Training records	X		X[e]	X				X										X		
Internal quality system audits	X													X		X				
Library control system[f]					X				X		X			X	X					

a. The chart assumes that the quality manager is responsible for scheduling and carrying out a system of planned and documented internal audits as required by the standards.

b. The project plan shows what skills are required and commits the necessary resources and skills to the task. Therefore, the project plan addresses the standards requirement that the supplier must have the capacity to meet contractual obligations.

continued

Figure 5.2. Essentials versus standards elements.

c. The chart assumes that the build plan is used for recording the test status of all parts, including incoming product. Therefore, the build plan addresses the standards requirement that where an incoming product is released for use before all incoming testing is completed, the product must be positively identified and recorded.

d. This chart assumes that the rationale and proof that a tool can serve its intended purpose is documented in the test plan for the project. The test plan addresses the standards requirement that test tools and equipment must be capable of serving the purpose for which they are used.

e. Training records are used to determine whether you have the skills needed to meet your contractual obligations (a requirement of the standards).

f. An automated library control system is the preferred mechanism for storing, controlling, and managing most information and data. There are a number of advantages to using a library control control system over less automated approaches. It also makes sense that not all information and data be kept in the library system. The standards elements marked in this chart are those where the library facility would be most beneficial. Again, most information and data are most often better kept in an automated facility.

Figure 5.2. (continued)

development. How you choose to implement the ISO 9000 standards requirements in your quality system is ultimately up to you.

Figure 5.3 depicts an ISO 9000 conforming quality system for software development. The figure incorporates the characteristics and items presented in this book along with the basic requirements stated in the ISO 9000 standards. It also indicates interactions among the various areas. In the figure, the product being developed (i.e., product items) includes parts that are developed internally, product documentation, and any included software or parts that are provided by subcontractors. The ISO 9000 standards require that these product items be controlled, identifiable, traceable, verified, and validated. These requirements can be satisfied through use of an ISO 9000 conforming quality system, which involves personnel, project-related items, and support items. The personnel need to be committed, involved, aware, and responsible. The project items must be controlled and should demonstrate the control, effectiveness, and auditability of the development process. The support items must be documented, effective, controlled, and continually improved. A quality system, made up of these items and exhibiting these characteristics, should provide the levels of control, discipline, order, and manageability necessary to develop software products that consistently meet the expectations of your customers.

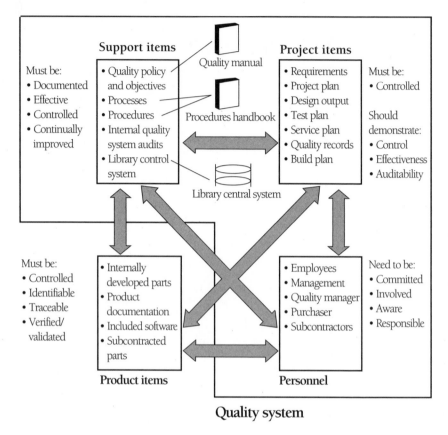

Figure 5.3. ISO 9000 conforming quality system for software development.

Notes

1. Throughout this book, written form can mean either hard copy or on-line, as long as the information is accessible, up-to-date, readable, and controlled. On-line, if possible, is preferable because it is easier to control.

2. Validation/verification involves more than testing. In software development, validation/verification can be accomplished in numerous ways, including reviews, inspections, walk-throughs, various levels of testing, and combinations of these.

3. This book makes a distinction between product documentation and internal documentation. Product documentation provides documentation of the product and is, therefore, an associated part of the final delivered product. Internal documentation is documentation associated with the development process, the quality system, and development of the product and is not part of the final product.

6

Steps to ISO 9000 Registration— Where to Start and How to Proceed

First, you have to *want* to register your quality system. You have to have made the decision that you are going to register your quality system. You must be committed to doing it and willing to expend the time, people, and money required to get it done. However, obtaining ISO 9000 registration should be secondary to your overall quality objectives and having an effective and fully functional quality system. You should have objectives that go beyond obtaining ISO 9000 registration; you should have quality goals. Producing quality products with a quality process should be of foremost interest. ISO 9000 registration is a necessary step, perhaps the first step, in ultimately getting to your quality objectives. Setting ISO 9000 registration as your only objective is likely to fail, if not initially, during subsequent surveillance audits. Remember, there are numerous benefits besides registration to having an ISO 9000 conforming quality system.

Once you have made the decision to seek ISO 9000 registration, you can follow the steps outlined below to get there. The sequence of steps assumes that you already have a process and a quality system you expect to register. If you are starting without an existing process, the order of these steps will be somewhat different.

1. Familiarize Yourself with the Standard.   Obtain the standard,[1] read it, understand it, and determine how it applies to your quality system and development process.

The standards are not lengthy or particularly difficult. They can, however, read like "legalese" with subtleties and hidden meanings. Applying them to your specific process (especially a software development process) can be challenging, sometimes perplexing, often frustrating, and somewhat time-consuming. You may want to hire a consultant to help you with this activity.

2. Determine the Level of Standards that Applies to Your Process.   Registration must be done against, and will eventually be granted to, one of the three levels of the ISO 9000 standards—ISO 9001, ISO 9002, or ISO 9003. You must determine to which level of the standards you should register. Many software

47

development organizations register to ISO 9001 because their development efforts include design activities or servicing after installation. For the purpose of this publication, ISO 9001 is assumed. ISO 9001 covers all stages of product development from design through servicing the delivered product. The level to which you choose to register should be commensurate with the development activities you normally perform. If you do design, then you should pursue ISO 9001 registration. Don't register to something less because you think it will be easier to attain. If your process includes design, register to ISO 9001. In this case, registration to ISO 9002 or ISO 9003 makes little sense and may give the impression that you are compromising your integrity for the sake of becoming registered.

Figure 1.4 explains the differences in the three levels of ISO 9000 and Figure 2.1 shows which of the 20 standards elements applies to which of the levels of the standard.

3. Assess Your Current Process Against the Standards. This step can be time-consuming and often proves to be most revealing. For those who have well-established, well-documented processes and procedures, this step can be relatively easy. An assessment of those processes and procedures by a person knowledgeable about the ISO 9000 standards is done. Often, however, processes and procedures are neither well established nor well documented. In fact, often, procedures are not documented at all. Sometimes procedures don't even exist! And there are always instances when the procedures that do exist (documented or not documented) are not being followed. All of these make it difficult to assess your current procedures since it first requires determining what the procedures are.

Once your processes and procedures are determined and documented, they should be assessed against the standard by a person knowledgeable in ISO 9000, in software, and in the software development process. Hiring a consultant for this step can be wise because consultants can add the objectivity of an independent party as well as provide interpretation of the standard to the software development process.

4. Identify What Needs to Be Done. After you have assessed your current process, you can identify what changes and additions need to be made to bring your quality system into conformance. Like the assessment step (step 3), this step requires a combined understanding of both the standard and the software development process because changes to the process will have to be integrated into the existing process to bring it into conformance without, at the same time, removing any existing conformance.

This can be tricky. It may require wholesale replacement of existing procedures. It may require simply adding some procedures. It may require integrating new procedures into the existing process. All need to be done in a way that brings the total process into conformance with the standard. The amount of time and difficulty involved in accomplishing this step depends on where and how severe the nonconformities are in the existing quality system. There are times when it may be necessary to establish an entire new process. Once the processes and procedures are established, they must be documented.

Consultants can help with this step; however, as much as possible, definition and documentation of processes and procedures should be done by your people—those who know them best and will use them.

5. Implement Changes.   You have now reached a critical decision point. You know what changes are required to become prepared for ISO 9000 registration. This may involve more than just process changes. It will require time and people devoted to making the identified changes. Time and people, of course, mean expense. It may require attitude changes and may even require some culture changes. The decision to continue toward registration requires commitment to that objective by all levels of management, willingness to allocate the required resources, and involvement by all of those who will be implementing and using the processes and procedures.

This stage needs to be well orchestrated. Fairly significant changes may be required and should be planned to be least disruptive. Your organization may be in the midst of a development cycle and making changes to the process while developing a product could be disastrous. It may not always be possible, and you may have to just stop the old way of doing things and start doing the new way. Most likely, you will wait until the current product development cycle is completed before instituting the changes.

6. Use Your Quality System. Your ISO 9000 conforming quality system needs to be proven. You need to use your quality system (i.e., it cannot just be a paper system) before the registration audit. You will need to show the auditors that your quality system is in operation and that it is effective (i.e., in the words of some auditors, it is a "mature" quality system). You cannot do this with a proposed system, it can only be done with a working, operational system. You might be able to show this with a system that has been operational for three months, but more realistically, expect six, nine, or 12 months of an operational quality system in order to satisfy yourself and the auditors that your quality system works and is effective.

7. Prepare for Registration.   Now that you have established and implemented an ISO 9000 conforming quality system, you need to make sure that you are ready to meet the requirements for registration. If you have not already done the following, you should do them now. They will absolutely be needed to pass a registration audit.

- Appoint a quality manager. Appoint a quality manager who is assigned the responsibility for getting your organization registered. Authorize the quality manager to make decisions and apply resources where needed. For consistency, it is a good idea to make the quality manager responsible for other aspects of the quality system as well, such as quality record ownership, quality manual ownership, quality system ownership, internal audits of the quality system, review of the quality system with management, identifying and implementing improvements in the quality system.
- Produce a quality manual, including quality policy. Although not required by the standards (revisions to the ISO 9000 standards, scheduled for late 1993, may make a quality manual mandatory), it is recommended that you produce a quality manual that provides an overview of your quality system, including, and usually beginning with, your quality policy. The quality manual is sometimes the only documentation auditors will review prior to an audit. Despite the fact that it has other uses, one of its main purposes is to convince the auditors that you have an operational ISO 9000 quality system. The quality manual should address each and every standards element pertinent to the level of standards for which you are seeking registration. For convenience and ease of reference, it is a good idea to organize your quality manual the same way the standard is organized. A quality manual can make it easier for an auditor to do the audit. For each standards element, make an overview statement describing, in sufficient detail to be convincing, how your quality system addresses the element. If the element does not apply in your development environment, say so and state why.

    Your quality policy should have been available by this time, but if it isn't, be sure to establish one now. It should be included in your quality manual.
- Document all relevant processes and procedures. Document all procedures that you would need documented if you had to replace all your people. See Chapter 5 for a discussion on the level of documentation required.

- Do a preassessment audit. Conduct a preassessment audit prior to the registration audit for your benefit. The purpose of this audit is to establish your readiness for registration and identify problems that would prevent you from gaining registration. Use it to convince yourself, as you would convince a registrar's auditor, that your quality system is ISO 9000 conforming, that it is being used, and that it is effective for you. This audit should be conducted in a fashion similar to the registration audit that will eventually be done. It can provide several valuable benefits.

  a. It will help you determine your readiness for an actual registration audit.
  b. It will identify problems in your quality system that would prevent you from gaining registration.
  c. It will provide you with an indication of how well your system is working and how effective it is.
  d. It will provide evidence to the auditors that your quality system is operational.
  e. It will satisfy the auditors that required internal audits have been implemented.
  f. It will provide experience to you and your people as to what to expect from the actual audit.
  g. It can help train your people on how to conduct internal audits, which are required by the standards.
  h. It will reemphasize the importance of your documented procedures to your own organization.

You can do your own preassessment audits, you can have your registrar do it, or you can have a consultant do it for you. If you do it yourself, it should be done by properly trained personnel. If the registrar does it, remember, the registrar can tell you *where* you do not conform, but cannot tell you *how* to fix nonconformities. This avoids conflicts of interest on the part of the registrar.

No matter who performs the preassessment audit, it is for you and for your benefit. It is intended to be constructive, to identify problems, and to point out weaknesses. You can choose to make it comprehensive or concentrate only on areas where you suspect your quality system may be weak. Whichever you choose will probably depend on how good (or bad) you feel about your preparedness at the time. Cost may also be a factor. The more comprehensive the audit, the more it will cost.

Some people believe that the more comprehensive the preassessment audit, the better. They believe that it is good to get a total assessment before the registration audit, enabling them to address and correct deficiencies prior to the registration audit; thereby leading to a smooth registration audit. Although this approach increases the level of confidence in attaining registration, it does not guarantee a successful registration audit.

Others believe that the preassessment audit should focus on weak areas and let the registration audit provide a more thorough assessment of their quality system. The danger with this approach is, of course, that the registration audit may uncover problems that delay your registration.

Having your registrar do your preassessment audit may help when the registrar does your registration audit because during preassessment the registrar becomes familiar with your organization and your quality system, its strong points and its weaknesses. Again, the registrar cannot suggest how to fix nonconformities but can only point out the nonconformities.

Finally, the registration audit is controlled entirely by the registrar. The registrar chooses what areas to audit, how comprehensive to audit, etc. So, again, having your registrar do your preassessment audit is not a guarantee that you will pass the registration audit.

- Select a registrar.[2] Third-party assessors, or registrars, are authorized by an accreditation body to conduct audits and to issue registration. In Europe, these accreditation bodies—e.g., National Accreditation Council for Certifying Bodies (NACCB) in the United Kingdom, Raad voor de Certification (RvC) in the Netherlands, and Ente Nazionile Italiano di Unificazione (UNI-CEI) in Italy—are governmental or quasi-governmental agencies. In the United States, the Registration Accreditation Board (RAB), an affiliate of the American Society for Quality Control (ASQC), regulates and accredits registrars.

  Accreditation of registrars is done to provide some level of quality and consistency among registrars, but, as of this writing, there is no single, international accrediting organization. As a result, there are some inconsistencies among registrars and some registrars may be better than others.

  In theory, ISO 9000 registration should have the same significance regardless of which registrar grants it. The registration scheme, however, is such that the certificate that is issued upon registration contains, among other things, the name of the registrar who granted the registration. Some people will judge the value of the certificate based on their opinion of the registrar (e.g., ISO 9000 registration by registrar ABC is better than ISO 9000 registration by registrar XYZ). This is unfortunate, but it happens. Several factors should be considered when choosing a registrar for the purpose of registering a software development organization.

a. Software skills. The registrar should have software skills or be knowledgeable in software and software development. You have to be able to communicate effectively with the auditors during the audit. Nothing is more frustrating than trying to explain what you are doing and having the auditor not understand (especially if the explanation pertains to a potential nonconformity). It is also important that the auditors be able to ask the right questions in fairly concise terminology. Software knowledge by the auditors is definitely beneficial. Because of recognized differences with software development, a separate, but similar, registration scheme, called TickIT, is currently being formulated for software (see Appendix I).

b. Schedule. Ask when the registrar can do the job and determine if it meets your schedule. The audit effort usually varies with the size of the organization being audited and, therefore, larger organizations will require more person-days of effort by the registrar.

c. Cost. Most registrars estimate the cost of the registration audit effort based on the size of the organization being audited. They usually charge on the amount of effort required (i.e., on a per person-day basis), but the fees can vary significantly from registrar to registrar.

Be sure to understand each registrar's postregistration requirements because you should figure these costs into your estimate as well. When comparing registrar costs, compare costs from one registration through reregistration (usually three years after the first). Understand the total registration-to-registration cost when you compare costs of registrars.

d. Accreditation. Until an internationally agreed upon accreditation scheme is established, you will want to know where the registrar is accredited. Usually each country has its own accrediting body, and many registrars are accredited in multiple countries (i.e., by multiple accreditation bodies). If, for example, you do much of your business in a particular country, you may consider using a registrar accredited in that country.

e. Post-registration requirements. The ISO 9000 series of standards does not specify third-party registration or audits. These are accepted ancillary activities that have been spawned as a result of the wide acceptance of the ISO 9000 series of standards. Periodic surveillance audits and reregistration are generally accepted practices but may vary both in frequency and rigor at the discretion of the registrar. Generally reregistration with a full registration audit is required every three years with surveillance audits every six months in between. Ask each registrar about its post-registration requirements.

f. Reputation. You may want to consider the reputation of the registrar. Some people will form opinions on how good your system really is based on the reputation of the registrar that granted your certificate of registration. You may want to consider how your registration will be viewed by the rest of the industry and by your customers.

After you've selected your registrar, you will be required to complete and submit an application to that registrar.

It should also be noted that hiring a registrar commits you, as the auditee, to a set of responsibilities that are outlined in *ISO 10011: Guidelines for auditing quality systems.* Briefly you are committed to the following:

—Inform your employees about the objectives and scope of the audit.

—Appoint responsible members on your staff to accompany members of the audit team (one for each auditor).

—Provide facilities for audit team use (e.g., room, desk, and telephones).

—Allow access to material and resources. Material that cannot be made available must be identified in advance of the audit.

—Cooperate with the auditors.

—Determine and initiate corrective actions based on audit findings.

- Prepare your organization. Proper preparation by your organization for the registration audit team visit can be beneficial to both you and the auditors. A cooperative and smoothly run audit can have a positive influence on the auditors. Remember also that the relationship with your registrar is likely to be a long-term relationship that does not end with the registration audit. So the start of the relationship should be as amiable as possible.

Every member of your organization should attend an awareness session that explains the ISO 9000 standards, your quality management system, what the audit is all about, and the benefits of registration. This session should be a refresher; it should not be the first time any of your people are hearing about ISO 9000.

- Make management aware. Make sure the management team is aware of the audit, its purpose, the scope, and the audit process. Also make sure management is aware of your quality policy. An uninformed and unaware management team can prove embarrassing during an audit and raise serious doubts in the minds of the auditors about your commitment to quality.

- Prepare your employees. Be sure that all affected employees are aware of the audit, what it is about, and that they may be involved in audit interviews. Instruct them that, if they are interviewed by an auditor, they should be open, honest, frank, cooperative, and courteous. Let them know that they need not be defensive or protective. Instruct them to answer "I don't know" when they don't know and "That's outside my area of responsibility" when something is outside their area of responsibility. They should not guess at answers. Stress that they should relax; they are not being evaluated, the quality system is.

  Make sure they answer the questions they are asked and not what they think they were asked. If they do not understand a question, have them ask for clarification. Instruct them not to volunteer more information than what is asked. Inform employees that auditors may take notes during the interview, and just because the auditor jots something down does not necessarily imply a problem.

- Appoint escorts. Auditors must be "escorted" during their visit. The escorts can play an important and influential role in the audit process. Assign as escorts members from your staff who know your company, your organization, your process, and your quality system. The main purpose for the escort is to guide the auditor from place to place, interview to interview, and to introduce the auditor to the interviewees. Basically their function is to make the auditor's movement at your site smooth and efficient. But the escort can be more than just a tour guide. Assigning a well-qualified escort can be a real benefit to you.

  Although escorts should not answer for others, they can help clarify misunderstandings or miscommunications between auditors and interviewees. Escorts should not become involved in the interviews unless asked for help by the auditor or the interviewee. Much of the value of the escort is gained between interviews when he or she has the opportunity to clarify or embellish some of the previous interviews. This type of help from the escort, if done unobtrusively, can have a very positive effect on some auditors. To other auditors, it can have a negative effect. Escorts need to be sensitive to this possibility and react accordingly.

On occasion, escorts may have the opportunity to direct an auditor to one person who is more knowledgeable about a subject than others. This can also make the audit go more smoothly, and it helps create a good impression.

• Prepare information for the audit team. An audit team will often need some information that is generally not available outside your organization, such as list of locations and sites, number of employees at each, nature of work at each, how work is divided among sites, organization charts, etc. Providing this information to the auditors is very helpful in their preparation.

• Arrange access to resources and facilities. Optimize the use of the auditors' time by ensuring that they will have ready access to people, places, records, and documents. An auditor's time spent searching and waiting for people, looking for records, seeking access to areas, etc., is wasted time—and you are paying for it. It also leaves a bad impression.

Good and thoughtful preparation for the registration audit will make a favorable impression on the audit team, will help the audit run smoothly, and can save you money by making the most effective use of everyone's time.

8. Perform a Registration Audit. Auditing is the technique used by registrars to assess a quality system for conformance to the ISO 9000 standard. Auditing consists of interviewing individuals in your organization as a means of determining if you are doing what you say you are doing. It is the auditor's way to "get to the system." Auditors will look for evidence of conformity to the standards and will report both conformities and nonconformities.

Auditors will interview people at all levels of your organization and you can almost be certain that they will talk to high-level managers to ascertain their understanding of your quality policy and their commitment to it.

Unlike the preassessment audit, the registration audit is dictated by the registrar's auditors. They decide whom to interview, which areas to visit, what documents to see, etc. Unless agreed otherwise earlier, the auditors must be granted access to the areas, people, documents, etc., that they request. These should be readily available to the auditors during the audit. It is frustrating to auditors (and possibly detrimental to the audit outcome) to have requested items continually unavailable.

You are required to provide each auditor with an escort during the audit. You can use this to your advantage. Assign knowledgeable escorts who can contribute to the success of your audit.

The registration audit is necessary to gaining ISO 9000 registration. You cannot become ISO 9000 registered without a registration audit. Although there is usually anxiety associated with any audit, if you have an ISO 9000 conforming quality system and have prepared properly, you should have little concern. The registration audit can provide an opportunity for you to show off your quality system!

9. Maintain Conformance After Registration.   Once you have achieved ISO 9000 registration, you must maintain and improve your quality system. The ISO 9000 series of standards requires continual review for improvement, implementation of improvements, and regularly scheduled internal audits be done by you.

Registrars require that they conduct continuing audits (i.e., surveillance audits) to ensure your continued conformance. Most registrars will plan two surveillance audits a year; these do not have to be scheduled in advance. Surveillance audits are usually less comprehensive than registration audits and may focus on a particular area or areas. The intent of surveillance audits is to satisfy the registrar that your quality system remains in conformance.

Full reregistration is required every three years. Reregistration requires a full registration audit by the registrar, but, given the periodic surveillance audits, this audit should present little problem.

The steps to ISO 9000 registration outlined in this chapter follow a logical pattern. The steps, however, do not necessarily have to be done in the stated order and some steps can be overlapped. Appendix F contains a condensed list of the steps described in this chapter for easy reference.

Finally you can expect that, if your quality system is ISO 9000 conforming, the auditors will recognize that. It is very unlikely that an audit team will find a quality system to be nonconforming when, in fact, it is conforming. Likewise if your quality system is not in conformance, the auditors will likely discover that. The best route to registration is to follow the logical steps outlined in this chapter and be sure that the quality system you are using conforms to ISO 9000. Do these things and the registration audit should present little problem.

Notes

1. Standards are copyrighted and must be obtained from their owner. ISO 9000 standards can be obtained from the American National Standards Institute (ANSI).
ANSI
Attention: Customer Service
11 West 42nd Street
New York, NY 10036
Phone: 212-642-4900

Copies of the U.S. counterparts to ISO 9000, ASQC Q90–Q94, can be obtained from the American Society for Quality Control (ASQC).
ASQC
611 East Wisconsin Avenue
P.O. Box 3005
Milwaukee, WI 53201
Phone: 800-248-1946 or 414-272-8575
Fax: 414-272-1734

2. A current list of registrars offering registration services to ISO 9000 as well as a list of U.S. and Canadian companies registered to an ISO 9000 standard are available from:
CEEM Information Services
10521 Braddock Road
Fairfax, VA 22032
Phone: 800-745-5565, ext. 430
Fax: 703-250-5313

7

About the Audit

A registration audit by an accredited, third-party registrar is required for ISO 9000 registration. You cannot get registered without it and, although there is a certain foreboding about any audit, the registration audit can be reduced to little more than a formality if you have an ISO 9000 conforming quality system and have done proper preparation. You can generally expect the truth to come out of a registration audit. If your quality system conforms to the ISO 9000 standards, it will be found to be in conformance; if it has major nonconformities, it will be found in nonconformance. The point is, don't expect to get a nonconforming quality system registered, but you can fully expect to get a conforming quality system registered.

Auditors have a way of going about their business which is quite effective. Because an audit most often consists of sampling (i.e., not every single aspect of your system is evaluated during the on-site audit), there can and will be oversights. However, skilled auditors generally are quite accurate with their findings. They look for conformities and report conformities and nonconformities element-by-element. It is unlikely that they will report nonconformities that do not exist. Auditors are more likely to miss a nonconformity than to report a nonconformity that does not exist. Most auditors are good and quite accurate in their assessment. Guidelines for auditors for conducting and managing an audit can be found in *ISO 10011: Guidelines for auditing quality system.*

When the registrar's audit team arrives, make the team feel welcome and as comfortable as possible. It is a good idea to meet with the team prior to the on-site visit just to introduce the key players (for example, meet for breakfast or for dinner the night before). This can help break the ice and put people more at ease.

Audit Formalities

The audit team visit will consist of three main activities: the opening meeting, the interviews, and the closing meeting.

The opening meeting is generally brief (less than an hour) and is used to introduce the audit team, to reconfirm the previously agreed to scope of the audit (i.e., ISO 9001, ISO 9002, or ISO 9003), and to state which areas the audit team plans to visit (remember, the audit team decides this for registration audits). The auditee can take the opportunity at the opening meeting to present any information which they feel is pertinent, informative, and helpful to the auditors and to introduce the escorts to the audit team.

The interviews with various managers and employees constitute the largest part of the audit activity and is the means by which the auditors assess your quality system for conformance to ISO 9000. (There will be more about auditing techniques later in this chapter.)

The closing meeting is also brief (less than an hour) and is where the audit team presents its findings. The presentation consists of brief introductory remarks summarizing areas of concern and/or complimenting especially good findings, followed by an element-by-element statement of conformance or nonconformance. Elements for which no nonconformities were found are often reported as, "No problems were identified in this area." Where nonconformities were observed or there was a lack of evidence of conformance, the report might read something like, "There is a lack of evidence that . . ." or "There is lack of conformance to . . ." If the auditor agrees that a particular element is not applicable to your operation, the element will simply be noted as "not applicable." After the element-by-element review, the audit team will state what its recommendation relative to registration will be. The audit team must report the results of the audit back to the team's company (i.e., the registrar) and make its recommendation relative to registration. The registrar company then officially informs the auditee as to the outcome of the registration audit.

Although the subject and seriousness of the closing meeting would dictate a formal meeting, the closing meeting is usually informal. The main reason for this is the limited amount of time that the audit team generally has to prepare. Normally the closing meeting is held at the end of a day spent auditing and, whether the audit takes one day or more than one day, the audit team often has as little as one-half hour prior to the closing meeting to prepare its findings for presentation. Therefore, presentations are usually handwritten and hastily put together. This, however, does not diminish in any way the validity of the findings.

Discussion can take place during the closing meeting but should not be allowed to turn into arguments. Observed nonconformities are noted and confirmed by the auditors at the time they are observed. At that time, the auditee has the opportunity to agree or challenge and discuss the observed nonconformity. At the closing meeting there should be no surprises. Discussions at the closing meeting are often brought up by managers who are hearing the findings for the first time, and the discussions often end quickly when an auditee employee or escort agrees with the auditor's finding. The meeting usually concludes with the auditee thanking the auditors for their efforts and committing to corrective actions. (More will be said about corrective actions in the next section.)

For audits that take more than one day, a wrap-up meeting is held at the end of each day to review the events of the day. The audit team reviews the areas visited, who was interviewed, and the team's findings, including nonconformities, for the day. Audits of more than one day have the advantage of providing some time (albeit, little) for the auditee to prepare for the audit outcome.

Levels of Nonconformities

Nonconformities normally fall into one of two categories: major and minor. A *minor nonconformity* is a single observed lapse of quality procedure requirement (e.g., lack of some documented procedures, inability to show how testing is valid). A *major nonconformity* relates to the absence or total breakdown of a quality procedure requirement (e.g., no evidence of testing, no internal audits, documented procedures/processes not being used). ISO 9000 registration will not be granted if a major nonconformity exists. Major nonconformities must be corrected and another registration audit conducted before registration will be granted. Some assessing organizations more realistically allow for a third category of nonconformance where several minor nonconformities are observed in the quality system.

Registration will normally be granted with minor nonconformities unless the nonconformities are so numerous or interrelated such as to indicate serious problems in the quality system. The audit team has to make a judgment regarding the severity of the combined minor nonconformities. If the team judges that the combined minor nonconformities do not constitute serious problems in the quality system, the team can still recommend registration.

If there are no major nonconformities, the auditee may commit at the closing meeting to correct the nonconformities within three months. In this case, registration generally will be recommended, and rectification of the nonconformities will be checked by the registrar perhaps over the telephone or during the first surveillance audit. On the other hand, when there are major nonconformities, registration cannot be granted, and a registration audit will have to be conducted again after the auditee has removed the reasons for the nonconformance.

Auditing Techniques

Skilled auditors have subtle but effective ways of getting the information they want. They will ask open-ended questions to get the interviewee talking and interject more pointed questions during the answer. It's relatively easy to get a person talking; just ask them what they do. People like to talk about what they do (and there are few people who are willing to listen!). Then while the interviewee is describing what he or she does, the auditor will ask questions, such as:

"Is that procedure documented?" "Can I see it?"
"Who provides that input?" "How do you validate it?"
"Do you do this by yourself?" "How do you interact with the others?"
"I don't suppose that is documented . . . is it?" (Be careful!)
"Who approved that?" "Can you show evidence?"
"How do you measure it?" "Who defines the measurements?" "Can I see them?"
"How is your progress monitored?"
"How did you evaluate the code you received from the subcontractor?" "Can you show evidence?"
"Where does it go after you are through with it?" (Auditors will often use answers to questions like this to determine who to interview next.)
"Who would be the owner of those records?"

Other auditing questions:

"Have you seen this document?" (the procedures descriptions)
"Are you familiar with internal audits?" "Have you been involved in any?"
"Are the results documented?" "Were any actions generated?" "Can I see them?"
"How long do you keep these records?" (Quality records must have a retention time—"forever" is not good enough.)
"How do you measure the quality of your product?" "How do you translate customer feedback on quality into action items?"
"Can you explain the process you follow to . . . ?"
"How is what you are building specified." "Do you have a copy?" "Can I see it?" (Here the auditor might look at the document for indications that it is obsolete, improperly paged, etc.). "Is this the most recent version?" "How do you know?"
"Where is unit test documented?" "How do you know the tests are adequate?"

"How do the managers interact?" (Question directed to a manager.)

"What kind of contracts do you deal with?" "Who are they with?" "Who approves them?" "How do you decide if it is a good contract?" "What are the acceptance criteria for delivery and contract fulfillment?"

"Do you evaluate the organization that will do the work?" "What are the acceptance criteria?" "Do you keep records of how good or bad the subcontractor is?"

"You provide to them and they provide to you. How do you assure that each is meeting their commitment?"

"I suppose none of how you do this is documented." (Cute!)

"By the way, do you keep records of this?" "For how long?" "Can I see them?"

"You produce the . . .What process do you follow to do that?" "Where is it documented?"

"How much of your process is documented?" (Be careful!)

Random Auditing Techniques

The following is a random set of techniques and approaches used by auditors. You may find them interesting and useful in preparing for dealing with auditors.

Auditors will often decide whom next to interview based on answers to "who" questions in the current interview. That is, they identify and follow a process chain through the organization or department. An audit team will divide the interviews by functional or organizational areas. For example, one auditor may interview in the test area while another auditor interviews in the design area. They will make these decisions based on the organization chart.

Auditors will more frequently use the word evaluate than the word test. Test to a software developer usually implies executing code, which is a somewhat limiting definition. Evaluate goes beyond executing code and can involve techniques such as desk checking, walk-throughs, etc. Executing code is not the only way to evaluate software.

Auditors are interested in how key decisions are made, recorded, and communicated. They may question more vigorously when key decisions are involved than when the decisions are less important. For example, a decision to make a design change after a project is into the implementation stage may be of more interest (or concern) to them than the decision to change a test case.

Auditors will attempt to get quick, honest answers from interviewees with statements and questions like:

"I suppose none of how you do this is documented."
"I don't suppose that is documented."
"For how long do you keep this?"
"How much of your process is documented?"

Auditors try to avoid interpreting the standard. Some auditors contend (arguably) that no interpretation is needed. *You* decide what needs to be done by your quality system for it to be *effective*. The auditor's approach is to ask you to demonstrate or show evidence that what you are doing in your quality system is effective.

Auditors will attempt to put the interviewee at ease. They ask questions without interrogating. They are not abrasive and they do not gloat. If they feel a particular interview is not going well or is going nowhere, they will end it and go to the next interview.

Auditors will seek confirmation of a nonconformity when they observe it, usually by asking directly. Also, if they suspect a nonconformity, they will dig for more evidence. They do this to avoid confrontations during the closing meeting when they present the observed nonconformities to the audited organization.

The audit team will need time and a private place to plan its approach, plot strategy, assign work, discuss and review findings, draw conclusions, and prepare for the closing meeting. A properly equipped office or room should be available for the audit team's use during the audit.

Summary

The registration audit is an assessment of your operational quality system for conformance to ISO 9000 standards. It is confirmation by an independent, qualified third-party that your quality system is indeed up to the standards. Audits are very effective when these techniques are used by skilled auditors who are knowledgeable in software and the software development process.

It is your responsibility to demonstrate that your quality system is effective for you. If you can do this, then you can avoid having auditors make judgments regarding how good or bad your quality system is.

Epilogue

Following are some final words of advice regarding the ISO 9000 approach to quality, quality systems, and quality systems registration.

Establish a quality policy that has management leadership, management commitment, and everybody's involvement. Your objectives should go beyond attaining ISO 9000 registration. Once you have attained registration, you should be looking to further improve your process, your quality system, and the quality of your product(s). Quality objectives are always needed and should be ambitious but achievable.

Control is key to satisfying the ISO 9000 standards. The product parts being developed as well as the processes, procedures, tools, and people that have an effect on the quality of the product must be controlled. If you can achieve this control, you have much of the ISO 9000 battle won and a good foundation for providing higher levels of quality.

Assess your quality system, implement necessary changes, and use your quality system prior to engaging a registrar for ISO 9000 registration. Do this and you should have little problem with registration or the registration process.

As much as possible, use your own people to describe, document, and implement your quality system. It may be necessary to use consultants to provide advice and guidance through and after preparation, but you have to do the actual implementation of your quality system and preparation for registration. It is your system, you are the owner, you will be its user, and you are responsible for it.

Prepare properly and registration should be incidental. Don't prepare just to put on a show for the auditors. Remember that it is your quality system that will be assessed. Prepare your quality system, your management, and your employees properly, and the ISO 9000 registration audit should be a breeze.

Monitor and improve your process and quality system to satisfy the ISO 9000 standards and to continually improve the quality of your product or service.

ISO 9000 provides an excellent foundation. Although ISO 9000 may not provide quality leadership or solve all of your quality problems, at a minimum it is considered a good business practice for a supplier company. It provides a base of good, technically sound process practices that can be used to consistently build quality products and to build further quality improvements.

ISO 9000 is to your advantage and benefit. Devoting whatever resources are required to implement an ISO 9000 conforming quality system will pay off eventually for you and your customers—often sooner than expected.

Appendix A

Terminology

The following are the words of the author and not necessarily official definitions (if "official" definitions exist). They are intended to add insight and clarity to some of the terminology associated with ISO 9000 and are used throughout this book.

Accreditation. Recognition by a regulatory board of an organization as being qualified to perform ISO 9000 assessment of quality system consistently, impartially, and competently. Accreditation ensures a level of quality and consistency among registrars.

Assess versus **audit.** An assessment of something is an evaluation of it. Auditing is a technique or methodology that can be used to do an assessment. In the ISO 9000 scheme, auditing is used to assess a quality system's conformance to the ISO 9000 standards. There are other ways (possibly more comprehensive) to assess a quality system against the ISO 9000 standards, such as reviews and walk-throughs. Auditing is a practical and effective technique used by registrars to assess quality system conformance to ISO 9000 standards.

Auditability. The ability to show that something exists, demonstrate what is being done, and show evidence of what was done.

Certification. The act of having your quality system assessed by an independent, accredited third-party for the purpose of confirming its conformance to the ISO 9000 series of standards and attesting to it in writing. Known as **registration** in the United States.

Configuration management versus **change control.** Configuration management is the procedure for the overall management of the entities being developed and includes the identification of the entities and constituent parts, tracking, and controlling changes to those entities and constituent parts. Change control is the procedure for controlling and managing changes to the entities and parts being developed and to the project. Change control could be considered a subset of configuration management.

67

Conformance versus **compliance**. Acting in accordance with the requirements of the standards. Usually people comply and things conform. The two words are used interchangeably throughout this book.

Contract. Specifies the obligations of all parties involved (i.e., the purchaser and the supplier). Obligations can include what the supplier will do (supplier requirements), what the purchaser will do, and what the product will do (product requirements).

Controlled versus **out of control**. In complex processes, it is sometimes difficult to establish all of what constitutes being controlled. Being controlled or under control is not being out of control, and, conversely, being out of control is not being controlled or under control. It is often easier to recognize when something is out of control than it is to determine that it is under control. If you consider any part of your process to be out of control (i.e., not under control), you must either add to or change your process to bring it under control. Regardless, the ISO 9000 standards require that you implement a quality system that keeps the product and its associated items under control.

Customer. The ultimate user, or recipient, of a product or service. Due to the contractual nature of the ISO 9000 standards, referred to as purchaser throughout the standards. Some claim that a customer is a purchaser who comes back a second time.

Developer. Organization responsible for producing and delivering a software product. May also design and service the product. In some cases, may also provide requirements for the product (i.e., product requirements). Due to the contractual nature of the ISO 9000 standards, it is referred to as supplier throughout the standards.

Development methodology. A set of mutually supportive and integrated processes and procedures organized into a series of phases that make up the development cycle of a software product.

Organization. Throughout this book, organization refers to the entity (e.g., a company, a subcontractor) responsible for producing products or providing services. Another meaning for organization is the management structure of a company. The ISO 9000 series of standards places no requirements on the management structure of a company.

Procedure. How you do the things you do. More specifically, how you perform the various tasks to achieve a result.

Process. The things you do. More specifically, the high-level sequence, or flow, of tasks that are performed during the production of a product or service.

Process versus **product.** ISO 8402 defines a product as "the result of activities or processes. It can be tangible or intangible, or a combination thereof." The ISO 9000 series of standards applies to the process by which a product is developed, not to the product itself. Products are not registered to ISO 9000 and products do not receive the ISO 9000 "stamp of approval." Processes (more precisely, quality systems) are assessed, and the organization using the assessed quality systems receives the ISO 9000 certificate of registration. You can be confident that products produced and services provided using ISO 9000 registered quality systems meet the specified requirements.

Production versus **development.** The ISO 9000 series of standards contains some terminology that has a clearer meaning in manufacturing environments than in software development environments. The word production is used in the standards to connote assembly or actual product manufacture. There does not appear to be a single, analogous stage in the software development process, although one could consider several activities to be comparable such as coding, code compilation/assembly, product integration, and copy onto tape or disk. Development, as in software development, includes everything from design, code/implementation, test, etc., through product shipment (i.e., activities associated with transforming requirements into deliverables and ultimately into a functioning software product). ISO 9000-3 defines development as "all activities to be carried out to create a software product."

Product documentation versus **internal documentation.** Product documentation documents the product and is an associated part of the final work product. Internal documentation provides information associated with the development process, the quality system, and development of the product that is kept in-house and is not part of the final product.

Purchaser. In a contractual situation, the party contracting to receive a product or service. More generally, the ultimate recipient of a product or service— the customer.

Quality assurance. All of the activities required to provide the assurance that the specified level of product quality is achieved.

Quality assurance system. See **quality systems**. Used synonymously with quality systems and **quality management system** throughout this book.

Quality control. Activities and techniques used to affect the quality of the product being developed.

Quality management. ISO 8402 defines quality management as "that aspect of the overall management function that determines and implements the quality policy."

Quality management system. See **quality systems**. Used synonymously with **quality assurance system** and quality system throughout this book. Quality management system is usually considered to be more encompassing than quality systems and includes such items as recruiting and employee well-being.

Quality product. A product that meets the customer's expectations. This definition of quality encompasses more than the absence of problems. Customers expect more than just a defect-free product; they have expectations regarding function, performance, price, size, timeliness to market, satisfaction with the product, etc. The formal definition for business from ISO 8402 is "the totality of features and characteristics of a product or service that bear on its ability to satisfy stated or implied needs."

Quality system. All of the procedures, processes, people, management, tools, and facilities which have an effect on building quality into a product or service. System implies some degree of synergy, smooth interaction, and transition among the constituent parts of the system. ISO 8402 defines quality system as "the organizational structure, responsibilities, procedures, processes, and resources for implementing quality management." See also **quality management system** and **quality assurance system**. The term quality system is used prominently throughout this book, but the three terms are used interchangeably.

The word process is used in the book when you might expect quality system or one of the other two terms. Process refers to the high level flow and interaction of tasks that are performed to produce a product. Process is used when the focus is more on the flow of tasks than the overall system. System refers to the totality of the elements that go into producing the product (e.g., process, procedures, plans, quality policy, and people).

Readily accessible, **necessary** information, **regularly** reviewed, **appropriate** review, **competent** personnel. Objective terminology such as this is used throughout the ISO 9000 series of standards. It is left to you to determine what is readily, necessary, regularly, appropriate, competent, etc., in your environment. Whatever you decide must be *effective* for you.

Registrar. An accredited, independent, third-party ISO 9000 assessor. Also known as **certifying body** or **notified body** in the United Kingdom and Europe.

Registration. The act of having your quality system assessed by an independent, accredited, third party for the purpose of confirming conformance to the ISO 9000 series of standards and attesting to it in writing. Also known as **certification** in the United Kingdom and Europe.

Subcontractor. The party from whom the supplier (i.e., the developer) obtains products or parts.

Supplier. In a contractual situation, the party contracted to provide a product or service. In terms of the ISO 9000 standards, the supplier is always the party whose quality system is under review. In software terminology, it is the product developer.

Traceability. The ability to show, at any time, where an item is, its status and where it has been.

Verification versus **validation.** Validation is the evaluation of a product against its specified requirements. Verification is the evaluation of the correctness of the output of various stages of the development process based on the exit criteria for that stage.

Appendix B

Quality Policy—Excerpts

The following are excerpts from quality policies to illustrate quality statements, objectives, and themes. This list is intended to give you some ideas to help you formulate a quality policy or trigger additional ideas with which to complement your existing policy. It is not intended to provide a multiple choice list from which to pick and use as your own quality policy. Your quality policy should seriously state your quality objectives and how you plan to achieve them. It should truly reflect your situation. Do not set quality goals that are incompatible with your organization's direction or that cannot be achieved.

A quality policy does not have to be lengthy or complex. A short, simple quality policy may be the most suitable for you.

All activities must be in compliance with the following policy:
Quality first and last
The customer is the main priority
Actions and decisions are to be based on facts and figures
Quality is considered from the onset, not just checked at the end
The aim is to do things right the first time.
Continuous quality improvement is our rule of life.
The aim is perfection.

Preference will be shown to suppliers that apply the principles of total quality.
Quality assurance is based on standard ISO 9000.
UCB, Chemical Sector, Belgium

Our customers expect faultless products and services. Our intention is to achieve this by preventive measures and continual improvement of our processes and the quality of services of our suppliers.
All employees and executives are brought into the process and obligated to make this quality policy a success.
Fulfillment of the requirements on a quality assurance system per ISO 9000 is a fundamental aspect of our quality policy.
IBM Manufacturing, Germany (translation)

73

IBM's market-driven quality goal is total customer satisfaction.

Delivering, in a timely manner, defect-free total solutions that are clearly superior to anything in the marketplace . . .

IBM Corporation

It is the policy of Corning Glass Works to achieve Total quality performance in meeting the requirements of external and internal customers. Total quality performance means understanding who the customer is, what the requirements are, and meeting those requirements, without error, on time, every time.

Quality is knowing what needs to be done, having the tools to do it right, then doing it right—the first time.

Corning Glass Works

Logica's policy is to produce and supply systems, software, hardware and consulting to high professional standards . . .

Logica plc

Rule #1: The customer is always right.

Rule #2: If the customer is ever wrong, reread rule #1.

Stew Leonard's, Norwalk and Danbury, Connecticut

Appendix C

Quality Manual—An Outline

The ISO 9000 series of standards requires that you ". . . establish and maintain a documented quality system . . ." It does not specifically require a quality manual. (Revisions to the ISO 9000 standards, scheduled for late 1993, may make a quality manual mandatory.) However, developing a document—call it a quality manual—that addresses each of the ISO 9000 standards elements can be helpful and is recommended. A quality manual, along with documented processes and procedures, can serve to document the quality system as required by the ISO 9000 standards. A quality manual can also be used for communicating your quality policy, goals, and objectives to all of your employees. If written carefully, it can also be used to advertise your quality policy and your quality system to prospective employees, customers, and prospective customers.

The content of a quality manual will vary from organization to organization. If quality manuals for two organizations that do things differently are the same, the quality manuals are probably not written in enough detail.

Here are several things to keep in mind when developing a quality manual.

1. Keep it simple. Everyone appreciates simplicity, and your quality manual need not be as complex as your quality system or product.

2. Be convincing. Don't just state what you do, but explain simply and convincingly how you do things.

3. Don't overelaborate. Don't say more than is necessary to be convincing, and don't say more than is necessary to serve the purpose of the document.

4. Keep it short. The quality manual does not have to be lengthy to be effective. More is not necessarily better.

About the Outline

This appendix contains an outline of a multipurpose quality manual. The outline provides a framework, along with some guidance, for what should go into a quality manual.

Although the outline is not based on a real quality system or on a single hypothetical quality system, there are several items that are assumed to exist and are referenced throughout the outline. However, each element is written independently from the others. Therefore, you cannot put all of the outline sections together and expect it to produce a cohesive quality manual. The items that are assumed throughout are as follows:

Handbook of control procedures—a controlled document that contains control procedures.

Handbook of assurance procedures—a controlled document that contains assurance procedures.

Project test plan—a controlled document that specifies the quality activities for the project, how the various verification, validation, evaluations, and testing will be carried out, why the various approaches were chosen, etc.

Project plan—a controlled document that describes the objectives of the project, the development process that will be followed, what will be done, who will do what, and when.

Build plan—a controlled document that identifies the constituent parts making up the product, plan for each, who is responsible, interdependencies, specifications supported by each, status of each, etc.

Service manual—a controlled document that specifies what maintenance activities will be carried out after the product ships, who is responsible for what, etc.

Measurement handbook—a controlled document that specifies the various measures and metrics that will be used during and after the development to determine the effectiveness of the quality system and the quality of the product. The rationale for each metric will be provided in the measurement handbook.

In addition, the following organizations and facilities are also assumed to exist.

Project manager—the manager with overall responsibility for the product under development.

Project management team—the managers under the direction of the project manager who have various responsibilities for the product under development.

Corporate quality assurance—the group with overall assurance responsibility within the corporation and that owns the quality system, the documented procedures (i.e., handbook of control procedures and handbook of assurance procedures), and the quality records.

Local quality assurance—the group with particular responsibility for the quality assurance of this project.

Quality manager—the manager who, among other things, is responsible for the quality manual.

Library control system—a set of automated facilities for storage, management, and control of the various parts under development and the associated project items such as handbook for control procedures, handbook for assurance procedures, build plan, and project quality plan.

The outline suggests a quality manual with five sections.

1. Quality policy

2. Introduction to the quality manual

3. The organization and its products

4. Quality system description

5. ISO 9000 element-by-element description

The Outline

Quality Policy. State your quality policy. Having it signed or endorsed by a top-level manager or executive is helpful.

Introduction to the Quality Manual. State the purpose, intended audience and expected use of your quality manual. This section should also state who owns the quality manual and give its revision history. This is also a good place to list all of the pertinent standards used by your organization.

Example
The purpose of this document is to accomplish the following:

State our quality policy.
Describe our organization and what we produce.
Describe our quality system.
Describe how our quality system addresses the elements of ISO 9000.

This document is available to all interested parties. Some of the more detailed documents may not be available outside of the XYZ company. This quality manual is owned by the quality manager of the XYZ company, who has final authority relative to its content and distribution.

Comments
State the owner in terms of title, rather than name, to avoid having to change the document every time the person in the position changes.

This example assumes there is a quality manager who, among other things, is responsible for the quality manual.

The Organization and Products.   Provide a high-level overview of your company and organization and the products or service it provides.

Example
ABC is the organization within the XYZ company responsible for the development of software in support of the hardware produced by the MNO organization in the XYZ company. This software is produced for the market, in general, in support of the hardware and the software requirements are provided to ABC by the product planning organization of the XYZ company.

Comments
Provide enough information to be informative without providing classified information such as number of employees. Try to provide information at a level that will not require frequent change.
    This example assumes that the ABC organization is responsible for providing software to support the hardware produced by another organization (i.e., the MNO organization) within the same company (i.e., the XYZ company).

Quality System Description.   Provide a description of your quality system, the basic process flow, and the product life cycle, or development methodology, used.

Example
Because of the size and complexity of most of our software development projects, our unmodified quality system applies to most projects. For the occasional project for which modifications to the quality system are required, the project manager documents the required changes and has them approved by the quality manager prior to beginning development.

Our quality system is based on the principles that all work items are under strict control and have owners with final authority and that all procedures affecting more than one person or are not intuitive to the person who will perform them are documented.

Comments
This description should not attempt to provide a detailed description of your quality system but more of the philosophy or basic principles behind your quality system.

This example assumes that most projects in this organization are similar in size and complexity and, therefore, the same process is used. However, the project manager can request variations to the process or even a different process if he or she is convinced that, due to the nature of the project, the changed process would be more effective.

ISO 9000 Element-by-Element Description.   Describe how your quality system addresses each element of the ISO 9000 standards. If an element is not applicable to what you do, state that it is not applicable and why.

Following, for each standards element, the outline provides (1) a brief description of what needs to be stated, (2) an example of how to state it, and (3) other comments.

4.1 Management responsibility

4.1.1 Quality policy. State that you have a quality policy and that it is given in the first section of this quality manual.

Comments
See Appendix B for excerpts from quality policy statements. Quality policies can be very eloquent or very brief and simple. It should represent your honest, achievable intention.

4.1.2 Organization. Provide a list of organizations or departments that are responsible for any aspects of your quality system or quality of the product you produce. Identify the responsibilities, authority, and interaction of each.

Example

Corporate quality assurance is responsible for the independent review of the quality system and controlling and maintaining the quality system. Corporate quality assurance is also responsible for performing internal audits of the quality system to determine that it is being used and is effective. Corporate quality assurance schedules, conducts, and reviews the results of internal audits and surveillance audits annually with management. It also owns the written procedures and is responsible for maintenance of the documented procedures. Once a year, corporate quality assurance is audited by a group designated by the senior executive of the ABC organization.

Local quality assurance is responsible for monitoring and auditing projects and for assuring that projects are following documented procedures. Local quality assurance conducts a postmortem review of every project to determine the effectiveness of the process and identify improvements. Local quality assurance is also responsible for assuring that resulting action plans are implemented. Local quality assurance management is also responsible for assigning adequate resources, including properly trained personnel, to all tasks associated with each development project.

The project manager is responsible for the quality control of the project and must make provisions for quality control and the implementation of specified procedures. The project manager is responsible for the development and implementation of the project quality plan, which identifies the quality activities for the project, and the testing that is required for this project, and tracks the status of planned testing. The project manager is also responsible for ensuring that adequate resources, including properly trained personnel, are assigned to each of the tasks within the development project.

All employees directly contributing to the quality of the product, are responsible for understanding our quality policy, following documented procedures, and performing their job to the best of their ability.

Comments

This example assumes a corporate quality assurance department, a local quality assurance department, and a project manager, each with responsibilities stated in the example. It also assumes the existence of a project quality plan for the project which identifies the required testing and quality activities for the project.

4.2 Quality system. Provide an overview of your quality system and the principles upon which it is built.

Example
Our quality system is based on the principles of quality management; that is, management of quality through quality control and quality assurance. Quality management is the responsibility of our executive management. Quality control is the responsibility of each, and every, project manager, and quality assurance is the responsibility of our corporate quality assurance and local quality assurance departments. Our quality system consists of the following:

- Quality policy
- Procedures for control
- Procedures for assurance
- Quality and training records

Our quality policy specifies our goals and objectives for quality.

Our procedures for control specify the planning, control reporting procedures, and development methodology that are followed for all projects. These procedures and methodology are documented in our handbook of control procedures. The handbook of control procedures is owned by the manager of corporate quality assurance who is responsible for control of the document and its content.

Our procedures for assurance specify the monitoring, auditing, review, validation, and verification procedures that are followed for all projects. These procedures are documented in our handbook of assurance procedures. The handbook of assurance procedures is owned by the manager of corporate quality assurance who is responsible for control of the document and its content.

Quality and training records are the records that are kept to demonstrate the effectiveness of our quality system, the effectiveness of improvements, and the training records for our employees. Quality records are owned by the manager of corporate quality assurance, and training records for employees are kept by their managers as part of their personnel file.

All records and documentation are kept in the library control system, all have designated owners, and all come under defined change procedures. Access to various records is controlled by the library control system, and access authorization to various records is granted through the owner of the requested records.

The quality system is owned by the manager of corporate quality assurance, who has ultimate authority over change, access, content, etc.

Comments
This example assumes the existence of a handbook of control procedures and a handbook of assurance procedures, which contain documented procedures as stated in the example. It also assumes local and corporate quality assurance departments.

4.3 Contract review. State how you ensure that you have a clear, agreed upon understanding of what is expected from a project, whether the project is being done on a subcontract basis or is being given the product requirements from a source outside of your organization.

Example
There are documented procedures for reviewing and understanding what is expected from every project. These procedures call for meetings, joint reviews, technical exchanges, sign-offs, etc., between the project manager and the requestor to ensure mutual agreement on what the project is to produce. The project manager is responsible for obtaining the required sign-off prior to the beginning of the project. The procedures for contract review are documented in the handbook of assurance procedures.

Comments
If the project is being done on a contractual basis, this statement will also have to address how you ensure that there is mutual understanding of what is expected from the contract in addition to procedures for assuring there is an understanding of the product requirements.

This example assumes the existence of a handbook of assurance procedures that contains descriptions for, among other things, contract review and approval.

4.4 Design control. State how you control the design, and changes to the design, and how you verify that the design you produce meets the product requirements.

Example
Product design is the activity that transforms the product requirements into a specification from which code can be generated. Several levels of design (e.g., component design, module design) may be necessary depending upon the size and complexity of the product being developed. Procedures for the various design stages, who is involved, responsibilities, reviews, approvals, change, expected inputs, expected output, required skills, technical interfaces required, interactions, dependencies, required verification, etc., are documented and available in the handbook of control procedures.

A project plan is produced at the beginning of each project. The project plan specifies the objectives of the project, the project schedule, how the project will be managed, what organizations will produce what parts, what subcontractors will be involved and their role, the various stages or phases the project will go through, etc. Basically the project plan specifies all the planned development activities for the project. The project plan is updated as the project progresses to reflect required changes in planned activities and is reviewed and approved prior to beginning each new phase of development.

The project manager is responsible for the project plan, its review, and approval, and for assigning adequate resources, including properly trained personnel, to the planned activities.

Comment
This element applies only to ISO 9001. It is not applicable to ISO 9002 and ISO 9003 since design is not covered by ISO 9002 and ISO 9003.

This example assumes the existence of a handbook of control procedures which, among other things, contains the documented procedures for design and design control. It also assumes a project manager and a project plan for each project.

4.5 Document control. State how you control the development of, and changes to, the documentation associated with the project.

Example
The importance of strictly controlled documentation is recognized and procedures for document control are documented in the handbook of control procedures. Document control procedures specify procedures for review, approval, distribution, recall, change, retention, etc., for documents.

All documents that are related to the project and have an effect on the quality of the product are controlled by these procedures. A list of all documents that are under these procedures, and their owner is kept by the project manager. Every document under control of these procedures has a management owner who is responsible for changes made to the document and to decide when the document needs to be reissued due to the number or nature of the changes. The owner of each document has the authority to make final decisions on content and changes to the document he or she owns.

Comments
This example assumes the existence of the handbook of control procedures which, among other things, contains documented procedures for documentation and document control. It also assumes a project manager for the project who keeps a master list of documents under document control, their owners, and the current version of each.

4.6 Purchasing. State how you ensure that parts or products obtained from outside your organization are what you expect before using them.

Example
When products or subproducts developed outside the ABC organization are included in the final product or used in the production of the final product, these products or subproducts are subjected to verification upon receipt and to rigorous verification as part of the final product. Details of this verification and validation are as specified in the project test plan. Such parts are not used prior to successful verification upon receipt.

Subcontractors are evaluated based on the acceptability of the product or subproduct they produced. A list of acceptable subcontractors is maintained. Corporate quality assurance is responsible for maintaining the list of acceptable subcontractors and ensuring the integrity of the list. It is also responsible for establishing and maintaining the criteria for adding subcontractors to and removing subcontractors from this list. The process and criteria for evaluating subcontractors and maintaining the subcontractor list is documented in our handbook of assurance procedures.

Comments
This example addresses *4.6.2 Assessment of subcontractors*. If no parts are obtained from outside your organization, state that this element is not applicable.

This example assumes the existence of the project test plan which, among other things, contains the information stated in the example. It also assumes a handbook of assurance procedures which, among other things, contains documented procedures and criteria for evaluating subcontractors.

Example
When preparing to purchase products or services from an outside organization, the project manager is responsible for reviewing the purchase order to ensure accuracy and precision of the order data. Every effort is made to include a well-documented, well-controlled quality system for development as part of the subcontract requirements.

Comment
This example addresses *4.6.3 Purchasing data.*

4.7 Purchaser-supplied products. State how products obtained from outside your organization to be used as the basis for the final product are verified upon receipt and how they are treated after acceptance.

Example
When our product is built on a part or product obtained from outside, we treat the received part in the same manner as we treat parts received from subcontractors. Our contract with the provider of the part specifies the agreed-upon expectations. When the part is received, it is evaluated against those expectations. The procedures for the evaluation are documented in the project test plan. Upon acceptance, the part is treated as if it were produced by our organization. It is tested again as part of the final product during final verification. Procedures for final product verification are specified in the project test plan.

Comments
It is very important that you agree in advance with the provider on what you expect to get. Without an agreement on this, you cannot properly plan your project and your project will be out of control.

This example assumes the existence of a project test plan which, among other things, contains the information stated in this example.

4.8 Product identification and traceability. State how you control and keep track of the (many) parts and part versions that make up your product before, during, and after product delivery.

Example
The project manager is responsible for developing and maintaining a build plan for the project. The build plan is used to plan the parts that will compose the product, specify the order in which they come together, identify their inter-dependencies, monitor their actual progress and status during development, and, ultimately, serve as a parts list for the delivered product. The build plan is a controlled document and is subject to the document control procedures.

Storage, management, and control of parts and documents is provided by our library control system. Procedures for using the library control system are documented in the handbook of control procedures.

Comments
This example assumes the existence of a library control system that is used to store, manage, and control product parts and product documentation among other things. It also assumes the existence of a handbook of control procedures which contains information on the library control system, its capabilities, and procedures for using it.

4.9 Process control. State how you control your production process, including following documented work procedures, monitoring progress, approval of processes and tools, and criteria for quality.

Example
Our company recognizes the necessity of a well-controlled process for developing software products that meet our customers' expectations. Implementation of the product is carried out in a number of well-defined, well-controlled phases, each with prespecified criteria and expectations. Responsibility for ensuring successful completion of each of the implementation phases is assigned to a manager. Continued and close monitoring of compliance to documented procedures and of progress through the process is done by the project management team and the local quality assurance department. A phase is not completed until the output from the phase has been successfully verified against the prespecified expectations for that phase. The verification procedures are specified in the project test plan. Successful completion of each phase is recorded. Monitoring is achieved through (1) formal reviews at the end of each development stage, (2) management monitoring of daily progress and activities during development, and (3) process audits.

In addition, we use a number of mechanisms for ensuring control throughout development, many of which are established early in the project.

We identify, control, and track all software items, including product documentation, from project inception through product delivery and customer installation. To facilitate servicing the product, some items are tracked even after the product has been delivered and installed. Through the various process control mechanisms, we can determine, at any time during development, what items will be in the final product, status of each, its schedules, etc. Using these control mechanisms, we can trace each item to its origin, including the specification or requirement that it supports.

The control mechanisms used consist of:

1. Project plan—specifies the plan for the development activities for the project. It includes project objectives, development methodology, and phase definitions for the project, project schedules, and project responsibilities. The project plan is owned by the project manager, is established at the beginning of the project and is updated, reviewed, and approved at major checkpoints during the project.

2. Build plan—identifies the parts that will compose the final product, including product documentation, the order in which they are required to come together, and their interrelationships and interdependencies. The build plan is initially established during the design phase of the project. The build plan is owned by the project manager.

   When the product ships, the updated build plan serves as a bill of material and is used during product servicing.

3. Test plan—specifies the various levels of testing, verification, and validation that is required for the project. The test plan is initially established during the design phase of the project and is updated, reviewed, and approved, as appropriate, during the project. The test plan is owned by the project manager.

In addition, after each planned software item reaches a certain point in its development, changes are strictly controlled. Change control procedures are documented in the handbook of control procedures. Changes to items affecting the product (e.g., product requirements, product specifications, project plan, build plan, and test plan) are also strictly controlled. Control procedures for these items are documented in the handbook of control procedures.

A library control system is used to help manage, control, and safely store items being developed, items affecting the product, and process and procedures documentation. The library control system and how to use it are described in the handbook of control procedures. Procedures for the activities for various phases of product implementation are documented and available in the handbook of control procedures and the handbook of assurance procedures. Each procedure has an owner who reviews the procedure periodically, identifies problems and potential improvements, records review results, and initiates necessary follow-up activities.

Comments
By continued monitoring, in addition to satisfying element *4.9.1 General*, this process also satisfies element *4.9.2 Special processes*, which applies to processes for products where full verification cannot be done prior to delivery, and product deficiencies may not become apparent until after the product is in use. Per the ISO 9000 standards, special processes require "continuous monitoring and/or compliance with documented procedures."

This example assumes a local quality assurance department and project management team with responsibilities as stated in this example. It assumes the existence of a project plan, test plan, build plan, and a handbook of control procedures and a handbook of assurance procedures, which contain documented procedures for process control.

4.10 Inspection and testing. State your overall test strategy and the various levels of testing that you perform during the development effort. Also indicate where the required testing will be identified and documented.

Example
All evaluation, verification, validation, and test requirements are identified and described in the project test plan which is owned by and the responsibility of the project manager. All parts, not produced under control of this process are evaluated upon receipt for conformance with the agreed requirements. No part is used before this evaluation is successfully completed. The amount and type of evaluation required depends on the level of quality control provided by the providing organization and is specified in the project test plan.

Comments
This example addresses *4.10.1.1* under *4.10.1 Receiving inspection and testing.*

Example
If it becomes necessary to use an incoming part before it can be successfully evaluated, this situation is noted such that the part can easily be identified in

the eventuality of a failure. The final product will not be released until the planned incoming evaluation, as specified in the project test plan, is successfully completed.

Comments
This part of the example addresses *4.10.1.2* under *4.10.1 Receiving inspection and testing.*

Example
The necessity for in-process inspection and testing is identified in the project test plan. If specified by the project test plan, this testing is carried out under the same controls as any other required testing. Nonconforming parts will be held until successfully completing inspection unless the nonconformance is judged to have no affect on the subsequent development activity. Nonconformances are recorded and corrected before the final product is released.

Comments
This part of the example addresses *4.10.2 In-process inspection and testing.*

Example
Numerous levels of testing are performed throughout the development process including testing of all incoming parts and extensive testing of the final product. The project test plan specifies testing required to assure that the final product conforms to the stated requirements. It also states the purpose of each level of testing. All testing specified in the project test plan is successfully completed prior to release of the product.

Comments
This part of the example addresses *4.10.3 Final inspection and test.*

Example
Record of successful completion of required testing is maintained by the manager responsible for the test and by the project manager. Local quality assurance monitors test records to determine progress and status of the project.

Comments
This part of the example addresses *4.10.4 Inspection and test records.*
    This entire example assumes the existence of a project test plan, which contains the information stated in this example.

4.11 Inspection, measuring, and test equipment. State how you choose the tools that you use or how you validate that the tools you use are suitable for their stated purpose. Also state how you control the tools that you use.

Example
Tools used throughout the development, testing, packaging, and delivery of the product are chosen after an evaluation of their effectiveness to do the job for which they will be used has been done. All tools used during a project are evaluated for effectiveness at the end of the project. Tools used in subsequent projects are chosen from the list of effective tools from previous projects. Results of tool evaluations are kept by corporate quality assurance. Because they are subject to change, tools, like code and product documentation, are strictly controlled and are kept in the library control system. Tools are owned by the corporate quality assurance department.

Comments
Handling tools similarly to the way code, product specifications, and product documentation are handled will help prevent different people from using different versions of a tools. While it is not clear that this level of control is required by the standard, it makes sense to provide this level of control anyway, and it demonstrates that test tools are controlled.

This example assumes a corporate quality assurance department which, among other things, is responsible for the items stated in this example.

4.12 Inspection and test status. State how you keep the test status of the various product parts throughout the development process.

Example
The status and results of all reviews, inspections, evaluations, test cases, etc., are recorded and are accessible during the development process. The person responsible for the review, inspection, or test case files a written report of the results to management. The project manager is the owner of these records. Not all test case results will be recorded individually. Some may be recorded at a logical consolidation point. Test results that reflect the test status of an identifiable software item are kept in the build plan such that the test status of any item can be determined at any time via the build plan.

Comments
It makes sense to record the results of all testing, reviews, etc. This is information that will likely be used later in assessing the effectiveness of your process. Note also that, due to the nature of most software product development, pure ratios of complete versus not complete test cases/inspections provide little indication of how far the product has progressed. Where the product stands relative to completion is better determined by where in the development process the product is.

This example assumes the existence of a project manager who is responsible for keeping the test status of all parts being developed. It also assumes that the test status for each part is recorded and associated with the build plan, allowing for easy access on a per part basis.

4.13 Control of nonconforming product. State what you do with products when defects are discovered.

Example
Defects can be discovered in a product either before it is delivered to customers or after it is delivered. Procedures for handling the product when a defect is discovered after the product has been delivered are addressed in *4.19 Servicing*, of this document.

When a defect is discovered before the product is delivered, the practice is to rework the product until it successfully completes final validation. No product will be shipped that does not pass final validation test. The rework process, which includes finding the problem, generating and applying fixes, and retesting, is called debugging. Debugging is planned for and continues until the product successfully completes all of its testing. Regression testing is also done, as appropriate, as part of the rework process.

When the product reaches a certain level of development, changes to the product—even to fix defects—are strictly controlled. Once the product reaches this level, documented procedures for change, including reporting and logging of problems, are followed.

Procedures for regression testing, validation testing, etc., are documented in the handbook of assurance procedures. Procedures for handling the product when defects are discovered prior to product delivery are documented in the handbook of control procedures.

Comments
This example addresses *4.13 Control of nonconforming product*. It assumes the existence of a handbook of assurance procedures and a handbook of control procedures.

Example
In the eventuality that successful completion of testing appears unlikely, a decision regarding the disposition of the product (e.g., abandon the product, ship with known problems, etc.) will be made based on a number of factors. If it is decided to ship the product with known nonconformities, the nonconformities will be documented and customers will be so informed. The authority to ship a nonconforming product lies with the highest executive in the organization and is based on advice from the product manager.

Comments
This example addresses *4.13.1 Nonconformity review and disposition.*

4.14 Corrective action. State how you investigate causes of nonconforming products and ensure that corrective actions take place to prevent recurrences.

Example
At the completion of every project, two major reviews are held: a process postmortem and causal analysis. A process postmortem is conducted by the corporate quality assurance department (the process owner) to review quality records and understand what went right, what went wrong, improvements that can be made, and the effectiveness of the quality system. Identified actions are documented and implemented. Results are documented and are included in the annual management review. Causal analysis is conducted by the product manager to review and understand the cause of nonconformities in the product and how to avoid them in the future. Identified actions are documented and implemented. The corporate quality assurance department is responsible for following up on the action items.

Comments
Some people use this section to describe mechanisms for recording and tracking problems in the product after it has been shipped. That's fine, but it is not the intent of this element of the standards. The intent of this element is to ensure that the process is being reviewed to identify process improvements and for the removal of quality system problems.

   This example assumes a corporate quality assurance department that owns the processes and is responsible for the items stated in this example.

4.15 Handling, storage, packaging, and delivery. State how you protect the product during development and ensure that your customers receive what you intend to send to them.

Example
While the software product is under development, the various parts are kept, stored, and maintained in the library control system, which provides extensive recovery, backup, and security procedures to protect the parts from being intentionally or unintentionally lost, destroyed, or stolen. Test cases and associated documentation are also kept in the library control system with similar recovery and back-up capabilities. The library control system and how to use it are documented in the handbook of control procedures.

Delivery of the product from the development organization is done by electronic media, which, if lost or damaged, can be reproduced from back-up information stored in the library control system.

A separate validation program is sent to each customer, which can be used to determine if the product received is what we had intended to send.

Comments
This example addresses *4.15.1 General, 4.15.2 Handling, 4.15.3 Storage, 4.15.4 Packaging,* and *4.15.5 Delivery.*

This example assumes the existence and use of a library control system for the storage and control of the items stated in this example. It also assumes the existence of a validation program, which is produced by the development organization and is provided to customers for the purpose of validating the content of the delivered software product.

This example also assumes that the library control system and how to use it are documented in the handbook of control procedures.

4.16 Quality records. State how you identify, record, and keep quality records.

Example
The quality records to be kept for each project are identified by the product manager in consultation with corporate and local quality assurance management at the start of the project. Quality records are necessary to demonstrate that the quality system is operational, being used, and effective.

Quality records are created, stored, and maintained for the purpose of demonstrating the achievement of the required product quality and the effectiveness of the quality system. These records are stored in the library control system with extensive recovery, back-up, and security capabilities to prevent either intentional or unintentional loss, damage, or theft. Quality records are retained through the completion of two subsequent projects, or two years, whichever is longer, for the purpose of quality comparison and tracking. After fulfilling their need for a project, the quality records are turned over to the corporate quality assurance department.

Comments
Be sure to specify an amount of retention time. The standard requires a retention time be designated, and "forever" is not sufficient. Because they do not change after they are created, quality records do not require change procedures.

This example assumes the existence of a library control system for storage of, among other things, quality records. It also assumes a corporate quality assurance department.

4.17 Internal quality audit. State how you plan to carry out internal audits of your quality system by qualified personnel for the purpose of (1) assessing that procedures conform to ISO 9000 standards, (2) ensuring that procedures are being followed, and (3) demonstrating that your quality system is effective.

Example
An internal quality audit is conducted every year to ensure that our documented quality system continues to conform to ISO 9000 standard and that documented procedures are being followed and are effective. It also determines if previously identified action items are being implemented. Audits are conducted more frequently if deemed necessary due to the number and severity of problems. The manager of corporate quality assurance is responsible for making this decision.

Results of the audits are reviewed, recorded, and action items identified. Audit records are reviewed as part of the management review of the quality system. The audits are performed by trained auditors and are the responsibility of the corporate quality assurance department. Procedures for the internal quality audits are documented in the handbook of assurance procedures.

Comments
This example assumes a corporate quality assurance department responsible for, among other things, scheduling and conducting internal audits. It also assumes the existence of a handbook of assurance procedures, which contains procedures for internal quality audits.

4.18 Training. State how you identify the training needs of your employees and how you satisfy those needs.

Example
An education/training plan, which is updated at least once a year, is kept for every employee. Every manager is responsible for identifying the training/education needs for each of his or her employees and then for planning and ensuring that those needs are satisfied. Part of making the determination of needs is to match the training/skills requirement of the job or task to which the employee is to be assigned with the employee's skills. Skills required in each job or task are listed in the documented procedures for the task and the skills possessed by each employee is kept with their personnel records. Employee records are kept by the employee's manager who is responsible for the accuracy and currency of the records. Documented procedures, which contains the required skills for the procedure, are owned and updated regularly by the corporate quality assurance department.

Comments
Although not specifically required by the standard, it is a good idea to systematically evaluate the effectiveness of the training/education that employees receive. For example, if an employee attends a class and the class turns out to be something other than what was expected, then the training requirement is really not satisfied. Actions should be taken to correct this situation—either change the class or obtain the training in some other manner. Likewise, if an employee does not perform satisfactorily in a class, then that class should not be counted in the employee's skill inventory. If you have instituted a systematic evaluation of employee training/education, then statements to this effect can also be included in this section of your quality manual.

To meet the requirement that records be readily accessible, it is advisable to keep training records with, but separate from, other personnel information.

This example assumes a corporate quality assurance department responsible for documented procedures.

4.19 Servicing. State how you service the product.

Example
Procedures that must be followed when providing service for the product are documented in the service manual. The service manual is owned and maintained by the manager of the service department. Service procedures include procedures for receiving, analyzing, assigning priorities and severity codes; routing problem reports; notifying the customer of the receipt of a problem report; and providing a target fix date for problem fixes. They also include procedures for notifying the proper departments within the ABC organization of the problem, internal reporting, monitoring and tracking, creating fixes and testing, mechanisms for providing temporary and permanent fix, etc. Records of maintenance activities are kept for later use in determining the effectiveness of the service process.

Comments
If service for your product is provided by a totally separate organization from the development organization, you may want to state that *4.19 Servicing*, is outside the scope of your quality system, and therefore, not applicable. Be aware, though, that your service organization may have to obtain its own ISO 9000 registration for your ISO 9000 registration to make sense.

The example for this section assumes the existence of a service department and a service manual that contains written description of the service process and procedures as described in the example.

4.20 Statistical techniques. State if you use statistical techniques (e.g., metrics) within and/or about your quality system and how you determine that the statistical techniques you use are valid.

Example
Various measurement, metrics, and statistics are used throughout our development process for such purposes as projecting the number of valid problems injected into the product during development, estimating performance, determining effectiveness of the process, etc. All measurements used are documented in the measurement handbook where each metric is explained. Each explanation describes the purpose for the metric, how and from where the necessary data are obtained, and the rationale for, and verification of, the metric. The measurement handbook is owned by the product manager who is responsible for ensuring control over the handbook and its content.

Comments
If you use statistical techniques, you should state that you verify the metrics that you use.

This example assumes the existence of a measurement handbook that contains descriptions of all metrics, statistics, and measurements used in the project. It also contains the rationale for using the statistics and the verification that the statistics used serve their intended purpose and are correct.

Appendix D

Procedure Description—A Template

The ISO 9000 series of standards requires that processes and procedures be documented. The standards also require that your development process be controlled. Your documented procedures should provide evidence that your process is under control. That is, the reader of your documented procedures should be able to tell that your process is controlled. When documenting procedures, follow a preestablished format for each procedure description—a format that indicates control. Documenting each procedure this way will help you identify procedures that lack control or need work. When the documentation is complete, it can be used to help convince an auditor that your process is controlled.

Keep your documented processes and procedures together in a process notebook, or data file, where they are well organized and easily accessible. Have a lead-in section that provides an overview of your entire process, explains the life cycle and development methodology being used, shows the overall process flow, and cross-references the various documented procedures. Because processes and procedures are subject to change, the process notebook must have an owner and be controlled.

For larger products and more involved processes, it may be necessary to provide several levels of process descriptions (e.g., process, subprocesses, procedures, etc.).

Following is a template for describing procedures. Topics within a procedure description that you cannot adequately describe may indicate that control is lacking in that procedure. Careful examination may show that improvement needs to be made to the procedure.

Procedure Name.   Provide a name for the procedure; for example, design, change procedure, component design procedure, or code review procedure.

Procedure Purpose.   State the purpose for the procedure—what is accomplished by the procedure.

99

Example
The purpose of XYZ test is to verify the functions specified in the programming specifications, including external user interfaces, hardware/software interfaces, application program interfaces, and error messages.

Procedure Owner. Name the owner of the procedure—the person who has ultimate responsibility for approving changes to or deviations from this procedure.

Interactions with Other Procedures. State how this procedure fits into the overall scheme of things within the development process and how it interacts with other procedures.

Example
The XYZ test is the final test phase before the product is turned over to quality assurance for quality certification. The test requirements for XYZ test are specified in the project test plan. All code has been previously tested at the component level, as specified by the project test plan and integrated by the build department into a single product for testing.

Procedure Input. State what the input to this procedure will be, from where it will come, the acceptance criteria for the input, who verifies the acceptability of the input, and who finally approves accepting the input.

Procedure Output. State what the output from this procedure will be, where it will go, the exit criteria for the output, who verifies the acceptability of the output, and who has final approval of its acceptability.

Activity Flow. If there are multiple activities within this procedure, specify the overall flow of these activities and how they interact with each other.

Activity 1 through Activity n. Describe what is done and how it is done for each activity within the procedure. Additionally, specify the following:

- Any special conditions under which the activity is done (or not done) and who decides
- What tools and support processes are used
- What rules are followed
- What practices and methodologies are used

- What conventions and standards are applied
- Dependencies and interactions with other activities and procedures
- The data to be collected and analyzed for future procedure improvements or for demonstrating process effectiveness
- The qualifications of those performing the activity

Change Procedure.   Outline the procedure to be followed to change this procedure. Since a change procedure may commonly apply to many procedures, including the change procedure itself, it may be documented once and referenced here.

Appendix E

Change Procedure—An Example

The following is a hypothetical change procedure description included here as an example of the format, content, etc., of a documented procedure. The procedure documented here is a valid procedure that, although written to describe a request for a change to the product design, could be used for requesting changes to other items. The procedure in this example assumes the existence of a programming specification, which is a controlled document that contains the output from the project design activities and is the document followed by those producing code.

Procedure Name. Design change procedure.

Procedure Purpose. The purpose of the design change procedure (DCP) is to ensure that changes to the programming specifications (PS) are managed and controlled to maintain product quality and development productivity. This procedure covers the following:

- Submission of a request for change to the PS. The request contains all the information necessary to enable a qualified investigator to make an informed recommendation about the proposed change.
- Actions required to make a decision on the proposed change.
- Ensuring that all items impacted by the proposed change are identified and changed.

The DCP is activated for a project as soon as the PS is ready for approval, at which time change requests can be submitted.

Procedure Owner. The DCP is owned by the manager of the product design department, this person has ultimate authority for changes to the procedure. The manager of the product design department is the final arbiter relative to all changes to the design for which there is not unanimity.

Interactions with Other Procedures. Changes to the PS as a result of proposed changes submitted through this procedure may impact other items such as code, publications, and testing requirements. Changes that result

103

from changes to the design must be identified and submitted through the change procedure in effect for the impacted items.

Procedure Input.   A design change request (DCR) is written and submitted to propose a change to the PS any time during the development process after the PS is ready for approval. DCRs can only be submitted to request changes to the design; they will not be accepted for proposed additional function. DCRs are submitted to the product design department and may be submitted by anyone.

Procedure Output.   The output from the DCP is a closed DCR. DCRs are closed with one of three dispositions.

1. Reject. Proposed change is rejected and is not recommended for future consideration.

2. Future. Proposed change is rejected for this release but should be considered for inclusion in some future release of the product.

3. Adopt. Proposed change is accepted for inclusion in this release of the product. Updates to the PS plus change requests for other impacted items must be made before the DCR is closed.

Activities in the Design Change Procedure

Create and Submit a DCR.   Fill out a DCR form requesting a change to the PS. Submit the DCR form to the project design department. Anyone can create and submit a DCR.

Assign an Investigator.   The change request coordinator reviews the DCR for completeness, assigns a number for tracking purposes, and assigns the DCR to an investigator for investigation. The change request coordinator should have one to two years of design experience.

Investigate.   The investigator reviews the DCR and proposed solution and recommends adopt, reject, or future. The investigator provides a written statement substantiating his or her recommendation. The DCR, along with the investigator recommendation and rationale, is forwarded to members of the DCR review board for its review and vote. The DCR investigator should be a senior staff member with two to four years of design experience.

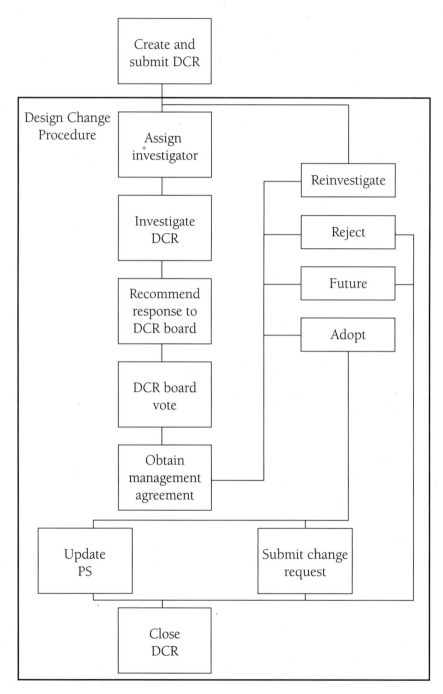

Figure E.1. Activity flow—design change procedure.

DCR Review Board Vote.   DCR review board members are responsible for reviewing each DCR and recommendation and returning their vote within five days of assignment of the DCR to the investigator. Meetings of the board members are usually not necessary to vote on DCRs unless there is disagreement on the vote. Votes must be unanimous. When they are not unanimous, a meeting may be required to resolve disagreements. The manager of the product design department is the final arbiter on all DCRs. The DCR review board should consist of qualified, currently assigned investigators.

Obtain Management Agreement.   The adopted DCR must receive management agreement from all affected and impacted item owners prior to being officially made part of the plan. Management agreement not only indicates conceptual agreement with the change, but indicates commitment to implement the changes.

Representatives from all impacted areas review the adopted DCR, assess the impact to their area, determine if they can accommodate the required changes, and make their recommendation to their manager. The manager, in turn, approves or disapproves the change. DCRs with agreement from all affected parties are incorporated into the plan. DCRs without total agreement must be escalated for resolution. The project design department is responsible for initiating the escalation.

Update PS.   When a DCR receives management agreement, the PS is updated by its owner. If sections of the PS have different owners, each owner is responsible for making the changes to his or her respective section.

Submit Change Request to Impacted Areas.   The project design department must submit change requests to each area impacted by the agreed design change.

Close DCR.   When the PS updates are complete and change requests to impacted areas are submitted, the DCR coordinator closes the adopted DCR. For rejected or future DCRs, the DCR coordinator closes the DCR as soon as the DCR review board vote is received.

The following is an example of a form that can be used for submitting and tracking a request for a design change. Note that this or a similar form could be used to request changes to other items as well.

Design Change Request (DCR) Form

Section I   (to be filled out by submitter)

Submitter name: name of submitter

Date submitted: date change request is submitted

Problem description: description of problem prompting proposed solution and change to design

Proposed solution: description of solution to problem causing the changes to design to be required

Functional areas affected by the proposed change: list all functional areas to be changed to incorporate proposed solution

Details of proposed change: provide additional detail about changes, if known

Other items impacted by the proposed change: list of items in addition to design, that will be impacted by this design change and how impacted

Additional comments: any additional comments important to the proposed change

Section II   (to be filled out by DCR coordinator)

Date received: date DCR received by DCR coordinator

Investigator: name of investigator assigned to investigate DCR

Date assigned to investigator: date DCR coordinator assigns DCR to investigator

Final disposition (adopt, reject, future, reinvestigate): final disposition of DCR as approved by management. When disposition is to adopt, commitments made by various managers when they agreed with DCR should be recorded here.

Final disposition date: date when management-approved final disposition decided

Date closed: date when DCR is closed. For adopted DCR, this is after PSs are updated and change requests submitted for other impacted items. Date closed may be considerably later than final disposition date.

Section III   (filled out by DCR investigator)

Date received: date DCR received by DCR investigator

Recommendation (adopt, reject, future): recommendation of DCR investigator. Note that investigator cannot recommend that DCR be reinvestigated.

Rationale for recommendation: investigator provides whatever rationale was used to make recommendation. This is important information to help DCR review board members decide on how to vote on DCR. This entry can serve as DCR investigator's report.

DCR review board vote: if vote is unanimous, may contain the final result (Adopt, Reject, Future, or Reinvestigate). If vote is not unanimous, may contain a tally of votes. Note that DCR review board can decide that DCR needs to be reinvestigated.

Date of vote: date when all votes received

Required action (as a result of the DCR review board vote): state actions to take place now that the DCR Review Board vote has been cast. Most interesting case is when vote is not unanimous and some resolution has to be achieved. This entry should specify how resolution will be sought.

Appendix F

Steps to ISO 9000 Registration

The following is a list of steps that should be performed in preparation for an independent third-party ISO 9000 audit. More information about each step is provided in Chapter 6.

1. Familiarize yourself with the standard. Get a feel for the standard, its applicability, and how it might impact your company.

2. Determine the level of standards that applies to your process. Select the one of the three levels of standard (i.e., ISO 9001, ISO 9002, or ISO 9003) to which you will target conformance.

3. Assess your current process against the standards. Review your documented procedures and processes and assess them against the level of standards to which you have chosen to comply. If your processes and procedures are not documented, document them.

4. Identify what needs to be done. Identify changes that will have to be made to your quality system to conform with your chosen level of the ISO 9000 standards.

5. Implement changes. Make the identified changes to your quality system.

6. Use your quality system. Make sure that your quality system is being used and that quality records, to enable you to demonstrate its effectiveness, are being kept.

7. Prepare for registration, include the following:

   • Appoint a quality manager.
   • Produce a quality manual.
   • Document all relevant processes and procedures.
   • Do a preassessment audit.
   • Select a registrar.
   • Prepare your organization.

- Make management aware.
- Prepare your employees.
- Appoint escorts.
- Prepare information for the audit team.
- Arrange access to resources and facilities.

Ensure that your organization has made the necessary preparations, including a preassessment audit, to facilitate a smooth and effective registration audit.

8. Perform a registration audit. Have an independent, third-party registrar conduct an official registration audit of your quality system.

9. Maintain conformance after registration. Correct problems identified during the registration audit and perform ongoing activities required by the ISO 9000 standards to maintain and improve your quality system. Remember, ISO 9000 registration is only a beginning!

Appendix G

ISO 9000 Self-Assessment

The following set of questions is intended to help you assess your readiness for an ISO 9000 registration audit. Questions are listed under the ISO 9000 standards element to which they apply. There is also a general category. The number in the parenthesis following the element name indicates to which of the three levels of the ISO 9000 standards the element applies.

The questions are written for a yes or no response (or NA if not applicable). For your quality system to be ISO 9000 conforming, you must be able to answer yes to *all* applicable questions. In some cases, a yes answer may evoke further questions, such as, "does that document have an owner?" Answer yes. Next question: "Who is it?" You should be able to answer all of these questions. Any questions to which you answer no should be investigated. Look for what needs to be done to change the no to a yes.

Questions preceded by an asterisk (*) are paraphrased questions extracted from the auditor guide found in the *TickIT Guide to Software Quality Management System Construction and Certification using EN 29001* (i.e., the TickIT Guide). These questions are included here because they may be asked by an auditor who is following the TickIT Guide. (See Appendices H and I for more information on ISO 9000-3 and TickIT.)

General (ISO 9001, ISO 9002, ISO 9003)

Is management aware of the audit? Its purpose? Management's role?
Is management familiar with your quality policy? Is management committed to it? Can you show evidence?
Are your employees aware of the audit? Do they know their roles?
Are your employees familiar with the procedures they use? Are their procedures documented? Do they know where? Are they following the documented procedures?
Do all of your processes and procedures have owners? Are the owners aware of the processes and procedures they own?
Is there a documented change procedure for all of your processes and procedures?
For each procedure, process, plan, and document, ask the following questions:

111

Is it documented?
Is there an owner? Is the owner aware of what items they own?
Is the owner's responsibility and authority defined?
Is the review and approval procedure documented?
Is a change procedure documented?
Is the distribution procedure documented?
Is there a documented procedure for disposal of obsolete documents?

4.1 Management responsibility (ISO 9001, ISO 9002, ISO 9003)

Do you have a quality policy?
Is it supported by your highest level of management?
Are your employees aware of it?
Do your employees have ready access to it?
Is your quality policy understood at all levels?
Is your organization committed to the quality policy?
Are resources available to implement the policy?
Do you have a quality manager?
Is someone responsible for ensuring that your quality system conforms
   with ISO 9000 standards?
Is your quality system reviewed regularly? Can you show evidence?
Is someone responsible to see that your quality system is reviewed?
Are the results of the quality system review reviewed with the highest
   level of management? Are action plans generated from these reviews?
Is someone responsible for their implementation?
 * Does your staff have defined responsibility and authority?

4.2 Quality system (ISO 9001, ISO 9002, ISO 9003)

Do you have a documented quality system? Is the documentation readily
   available?
Does your quality system have an owner? Is the owner aware that he or
   she owns it?
Does your quality system documentation have an owner?
Is your quality system approved by an authorized person?
Do you have documented procedures for changes to your quality system?
Are your processes and procedures documented?
Do you have documented procedures for changes to processes and
   procedures?

Do your processes and procedures have owners? Are the owners aware of the processes and procedures they own?
* Are there adequate procedures for documenting and controlling the quality activities for each product development on the basis of the quality system?
* Does your quality plan or equivalent clearly define quality objectives? The types of test verification and validation activities to be carried out? Detailed planning of test, verification, and validation activities? Specific responsibilities for quality activities?
* Is the documented quality system being adhered to?

4.3 Contract review (ISO 9001, ISO 9002)

If the project is being done under contract, do you review the contract with the purchaser to ensure you understand and agree on what is expected?

Do you review your product requirements with their originator to ensure that they are clear, understood, and what the originator intended?

Is there a procedure for this review? For problem resolution?

Do you have procedures for ensuring that product requirements are mutually understood and agreed?
* Do you have documented procedures for ensuring that the requirements are defined and understood by both you and the requester?
* Are there documented procedures for maintaining control of the requirements, including means of resolving differences of understanding and agreeing to changes?
* Are there documented procedures that establish what documentation is used for communicating requirement changes?
* Are responsibilities and interfaces clearly defined?
* Is consideration given to possible contingencies and risks associated with the contract, such as assessment of risk factors? Purchaser's capability to meet his or her obligation? Purchaser's role in requirements specifications, installation, and acceptance?
* Is there provision for an effective interface with the requester during design (e.g., to allow for acceptance of changes that arise from design planning)?
* Are requirements of and procedures for any acceptance tests addressed in the contract?

* Are service contracts reviewed?
* Do service contracts clearly define the type and levels of service provided? Service performance? Contact interface? Procedures to ensure confidentiality and protection of customer data and information during service activities? Duration of service provided? Clear statement of warranty?

4.4 Design control (ISO 9001)

Do you have documented procedures for your design process?
Do you have design review and approval procedures?
Do you have procedures for changes to the design?
Do you do design reviews?
Is the originator of the product requirements involved in design change review and approval?
Are the results recorded?
Are final design reviews held by independent parties (i.e., someone other than the designers)?
Before starting the design, do you have an initial plan specifying the various design and development activities and responsibilities? Are the development plans updated as the design evolves?
Do you assign design responsibilities to qualified personnel?
Do you have procedures for verifying that the result of the design activity will satisfy the product requirements?

* Do your procedures for controlling the design provide for the production and review of a detailed development plan before detailed design and development is started? The use of change control procedures? Design reviews to highlight and resolve outstanding technical issues? Holding progress reviews to ensure that adequate resource plans are available and resources assigned to permit effective execution of development plans? Reviewing suitability of the design methodology? The utilization of previous design experience?
* Are there procedures for reviewing and maintaining the development plan?
* Can you show evidence that all design and verification activities have been planned and assigned to staff with adequate training, experience, skills, and resources?
* Can you show evidence of effective operation of relevant organization and technical interfaces, especially in design review, product testing and acceptance, and product installation and support?

* Do you verify that the design input clearly and precisely describes each of the essential requirements of the hardware, software, and the external interfaces?
* Do you ensure that design input requirements are such that they can be objectively verified and validated by a prescribed method (e.g., inspection, analysis, demonstration, or test)?
* Is there a process for determining the design output criteria for each design input, including an assessment of whether standards are in place covering notation, structure, graphics, definitions, and measurability?
* Is your plan for design verification appropriate to the design activity to be verified? Are relevant employee skills described? Is it repeatable to the extent that equally qualified staff could obtain a similar result? Are the result and any further actions required to meet the design output criteria recorded and reviewed until the actions are completed? Are only verified design products released to the subsequent activity, unless a concession procedure is applied to the release of unverified products?
* Does your design verification examine the interfaces between software, hardware, and user documentation?
* Do your procedures for the identification, documentation, review, and approval of design changes address all design input requirements and design output criteria subject to change control? All verified design material subject to change control? Proposed changes subject to review at an appropriate level to ensure the early identification of impact on performance, sizing, cost, time scales, reverification, and documentary changes? A change note (or electronic equivalent) produced to record all aspects of the design change including authority for implementation? The timely notification to all relevant staff of impending changes? Traceability that exists between design change documentation and all items modified as a result of the change?

4.5 Document control (ISO 9001, ISO 9002, ISO 9003)

Do you have procedures for controlling your documents? Are there procedures for review? Approval? Distribution? Change? Withdrawal? Reissue?

Do all your documents have owners? Are the owners aware of the documents they own?

Do all documents have a means for determining their latest level? Their completeness? What changes have been made?

Do you have a master list that identifies the latest level of all your documents? Is the master list readily available?

Does the master list have an owner? Is the owner aware that he or she owns the master list?

* Are all documents concerning the product development subject to design documentation? Planning documentation? Procedural documentation? Reference documentation?

4.6 Purchasing (ISO 9001, ISO 9002)

Do you have procedures for verifying that parts or products obtained from outside your organization are what you expected?

Do you select subcontractors based on their ability to provide what you want?

Do you maintain a list of acceptable subcontractors?

Do you have criteria for putting or keeping subcontractors on the list? Removing subcontractors from the list?

* Are there procedures for the control of subcontracted software development?

* Are there procedures for controlling the activities of subcontract personnel employed on and off the premises to ensure that they meet subcontract requirements?

* Do you have sufficient data to allow verification and validation of the acquired product before using it?

4.7 Purchaser-supplied products (ISO 9001, ISO 9002)

Do you have procedures for verifying that parts or products received from outside your organization are what you expected?

* Do you have procedures covering the licensing and maintenance of software products supplied by the purchaser or by a third party?

* Do you control the purchaser-supplied product the same way you control products you produce?

4.8 Product identification and traceability (ISO 9001, ISO 9002, ISO 9003)

Do you have procedures for keeping track (i.e., where they are, latest level or version, status, etc.) of the various parts as they are being developed?

* Do you have effective procedures for configuration management of all hardware, software, and deliverable documents being developed, serviced, or used? In particular, do you have procedures for unique identification of the versions of each item? Identification of those versions of each item that constitutes a specific version of a complete product? Identification of all documents, development tools, and computer files for each version of a configuration item? Identification and tracking of all actions and changes resulting from a change request?

4.9 Process control (ISO 9001, ISO 9002)

Are your employees following documented procedures?
Do you have procedures for monitoring the progress of your projects?
* Do you have procedures detailing quality controls appropriate to all processes directly affecting quality?
* Are all software tools used in software development under configuration control? Identifiable by software product name and/or number, issue or version number, and date of issue? Recorded or registered, such that (in addition to the foregoing data) details of their development origin or source of supply are available?
* Do you have procedures for establishing confidence in external or proprietary software selected for use or for inclusion in your product?

4.10 Inspection and testing (ISO 9001, ISO 9002, ISO 9003)

Have you identified and documented the inspection and testing required for the project?
Can you demonstrate that the testing that you identified will adequately validate the product?
Do you have procedures for the required testing?
Do you maintain records of test activities?
Can you show evidence that the required testing has been done?
* Do you verify that inspection and test tools fit the purpose and are maintained in an agreed manner?
* Are test functions appropriately specified and controlled?
* Do procedures exist to ensure that test planning, execution, and recording is carried out appropriately on individual software items, integration of software and hardware system components, and the whole system?

* Is fault information recorded at all levels of testing?
* Are replication processes, if used, subject to inspection and test activities at appropriate stages?
* Are specific test responsibilities of the customer identified and documented?
* If a number of versions of a product have been released, are test records for each version retained?

4.11 Inspection, measuring, and test equipment (ISO 9001, ISO 9002, ISO 9003)

Can you demonstrate that your test tools are capable of verifying your product?
Are your test tools/test cases controlled such that you are sure everyone is working with the proper level?

4.12 Inspection and test status (ISO 9001, ISO 9002, ISO 9003)

At any point in time can you determine your test status? What has been done? What remains to be done?
* Do you have suitable means for identification of inspection and test status?

4.13 Control of nonconforming product (ISO 9001, ISO 9002, ISO 9003)

Do you have procedures for what to do when defective parts or products are discovered?
* Have you made provisions for the declaration of known nonconformity to your customers?

4.14 Corrective action (ISO 9001, ISO 9002)

Do you have procedures for determining the cause of nonconforming products?
Do you have procedures for identifying weaknesses in or areas for improving your quality system?
Can you show evidence that corrective actions have been identified?
Can you show evidence that corrective actions have been implemented?

* Do you have procedures for dealing with customer complaints and, in particular, that records of all complaints received and actions taken are retained?
* Do you have procedures for analysis of reported software faults to determine the cause and make provision for review of methods, tools, and processes employed in development?

4.15 Handling, storage, packaging, and delivery (ISO 9001, ISO 9002, ISO 9003)

Do you have procedures for safe, secure storage of the product parts throughout development?

Do you have adequate recovery procedures?

Do you have procedures for ensuring that your customer receives the product that you shipped (or thought you shipped)?

* Do you have provisions for maintaining the security and integrity of your product during development?
* Is the media being used to store the product controlled as prescribed by the manufacturer?
* Can the information be reconstituted in case of media damage or corruption?
* Is access to data controlled?
* Do you have procedures to ensure that the correct version(s) of software items are transmitted? Distribution is made only to the correct destinations?

4.16 Quality records (ISO 9001, ISO 9002, ISO 9003)

Have you identified the quality records you need to keep? Have you indicated retention time for each?

Are you satisfied that the identified records will enable you to demonstrate that your quality system is effective for you? Can you prove that?

* Have you made provisions so that pertinent records are identified? That they are kept for a sufficient time to enable effective support of the various versions of software items distributed to individual customer sites? That they contain sufficient data to facilitate the corrective action process?

4.17 Internal quality audit (ISO 9001, ISO 9002)

Do you have a documented plan and procedures for conducting regular
internal audits of your quality system?
Can you show the results from your last internal audit?
Can you show evidence of who reviewed the results? Planned actions?
Were the planned actions implemented?

4.18 Training (ISO 9001, ISO 9002, ISO 9003)

Do you know what skills or training is required to perform each procedure? Can you show it?
Do you know the skills and training of each of your employees? Can you
show it?
Can you show what training is required for each of your employees?
* Do you have procedures to identify the training needs and to provide for
the training of personnel that take into account the specific tools, techniques, methodologies, and computer resources? The specific field of
application of the product?
* Do you maintain records of training and related experience so that assignment of staff to work in all parts of the product life cycle can take such
training and experience into account?
* Has staff assigned to a specific task been trained according to the needs?

4.19 Servicing (ISO 9001)

Do you have procedures for providing service to your product after it has
been delivered to your customer?

4.20 Statistical techniques (ISO 9001, ISO 9002, ISO 9003)

Can you show what statistical techniques are used and for what purpose?
Can you show what metrics are used?
Can you prove that each statistical technique accomplishes its objective?
Can you show your methods for collecting statistics? Can you demon-
strate that they are correct and accurate?
* Do you have an adequate system for recording reported faults of software
during development, installation, and any period of warranty?

Appendix H

ISO 9000-3—Additional Guidance for Software

Overview

Recognition that the process for development and maintenance of software is different from that of most other types of industrial products coupled with some confusion and difficulty in applying the ISO 9000 standards to software have caused the ISO Technical Committee, ISO/TC 176, to prepare *ISO 9000-3: Guidelines for the application of ISO 9001 to the development, supply and maintenance of software* (referred to throughout as ISO 9000-3). ISO 9000-3 provides guidance on how to apply ISO 9001 to software development. The guidance is excellent and should be understood and given serious consideration when designing quality systems for software. However, ISO 9000-3 is a guideline and should be recognized as such. If you decide to do things differently from what is suggested by ISO 9000-3, you should at least be sure to understand how the alternative you chose addresses and satisfies the requirements of the standards.

Structure of ISO 9000-3

ISO 9000-3 is structured and organized based on the premise that associated with each software development project is a life cycle consisting of a set of phases, or defined segments of work. It assumes no particular life cycle, but it does assume the existence of some life cycle with phases.

ISO 9000-3 also presumes that the software product produced is the result of a contractual agreement between a purchaser and a supplier, where the supplier is the company or organization whose quality system is under review (i.e., ISO 9001 conforming). It emphasizes cooperation and joint efforts between the purchaser and the supplier, and there are clauses in ISO 9000-3 that specifically address the purchaser's responsibility in the contractual arrangement.

ISO 9000-3 consists of 22 clauses that do not correspond directly with the 20 clauses of ISO 9001. The 22 clauses of ISO 9000-3 are grouped into three major sections. Section 4, *Quality system—Framework*, deals with activities which are neither phase nor project related but persist across projects.

121

Section 5, *Quality system—Life-cycle activities*, deals with activities that are related to particular phases of the development process and are project related. Section 6, *Quality system—Supporting activities*, deals with activities that are not related to any particular phase but apply throughout the process. Figure H.1 shows the content of ISO 9000-3 and cross-reference between the clauses of ISO 9000-3 and the elements of ISO 9001 to which each applies. The source of this cross-reference is Annex A of ISO 9000-3.

It is important to understand the difference between ISO 9001—the standard, and ISO 9000-3—the guidelines. Certification and registration of a quality system is to the standard not to the guidelines. The guidelines offer

Clause in ISO 9000-3	Clause in ISO 9001
4.1 Management responsibility	4.1
4.2 Quality system	4.2
4.3 Internal quality system audits	4.17
4.4 Corrective action	4.14
5.2 Contract review	4.3
5.3 Purchaser's requirements specification	4.3, 4.4
5.4 Development planning	4.4
5.5 Quality planning	4.2, 4.4
5.6 Design and implementation	4.4, 4.9, 4.13
5.7 Testing and validation	4.4, 4.10, 4.11, 4.13
5.8 Accceptance	4.10, 4.15
5.9 Replication, delivery, and installation	4.10, 4.13, 4.15
5.10 Maintenance	4.13, 4.19
6.1 Configuration management	4.4, 4.5, 4.8, 4.12, 4.13
6.2 Document control	4.5
6.3 Quality records	4.16
6.4 Measurement	4.20
6.5 Rules, practices, and conventions	4.9, 4.11
6.6 Tools and techniques	4.9, 4.11
6.7 Purchasing	4.6
6.8 Included software product	4.7
6.9 Training	4.18

Figure H.1. Cross-reference ISO 9000-3 to ISO 9001 (Source: ISO 9000-3, Annex A.).

suggestions or recommendations about things to do or consider that will help a software development organization satisfy the requirements of the standards. In most cases there are other ways to satisfy the standards. You do not have to follow the suggestions made in the guidelines, but you should consider what aspects of the standards each addresses. Be sure that those aspects of the standards are being addressed by whatever you do use. Again, it is the standards that must be satisfied, not the guidelines.

TickIT auditors will use ISO 9000-3 as a guide and will look at your quality system for the items suggested in ISO 9000-3. If the auditors do not find these items, they will likely expect an explanation as to how the standards are being satisfied.

Guidance by ISO 9000-3

ISO 9000-3 offers additional guidance for software in the form of (1) *clarifications* of terms peculiar to software and software development, (2) *considerations* specifically for software, and (3) *items* to implement. There are several clauses in ISO 9001 for which ISO 9000-3 provides no additional guidance. Other than being numbered differently, the following clauses in ISO 9000-3 are unchanged from ISO 9001.

Clause in ISO 9000-3		Clause in ISO 9001
4.1.1.1	Quality policy	4.1.1
4.1.1.2.1	Responsibility and authority	4.1.2.1
4.1.1.2.1	Verification resources and personnel	4.1.2.2
4.1.1.2.3	Management representative	4.1.2.3
4.1.1.3	Management review	4.1.3
4.3	Internal quality system audits	4.17
4.4	Corrective action	4.14
6.2.4	Document changes	4.5.2
6.3	Quality records	4.16
6.7.2	Assessment of subcontractors	4.6.2

Clarifications
ISO 9000-3 provides the following clarifications of terms as they apply to software.

Maintenance. Clauses 5.10.1 and 5.10.5 of ISO 9000-3 classify the following as maintenance.

- Problem resolution; correction of software nonconformities (i.e., "bugs")
- Interface modifications; software changes to accommodate changes to the hardware
- Functional expansion or performance improvement; software changes to expand or improve an existing function or to improve performance

Design and implementation. Clause 5.6.1 of ISO 9000-3 describes design and implementation activities as ". . . those which transform the purchaser's requirements specification into a software product."

Configuration management. Clause 6.1.1 of ISO 9000-3 describes configuration management as "a mechanism for identifying, controlling, and tracking the versions of each software item."

Purchased product. Note 7 in clause 6.7.1 of ISO 9000-3 describes a purchased product as a ". . . software and/or hardware item intended for inclusion in the required end product or a tool intended to assist in the development of the required product." [Notice that ISO 9001 element 4.6 *Purchasing* requires that the supplier ensure that purchased products conform to specified requirements. Using the above ISO 9000-3 description of "purchased product," the supplier is required to ensure that any purchased product, including those used in the development of the product (e.g., compilers, project management tools, library control, and data base systems) meet their specified requirements.]

Items

The following is a set of items, along with a brief description, suggested by ISO 9000-3 to be implemented as part of the quality system.

Quality plan. Clauses 4.2.3 and 5.5.1 of ISO 9000-3 call for a quality plan that itemizes the various quality activities that will be undertaken for the project. The quality plan should specify quality objectives in measurable terms, input and output criteria for each development phase, identification of the types of tests, verification and validation activities for the project, schedules and resources for various test activities, responsibilities, and how defects will be controlled and corrected. A quality plan would be produced for each project.

Requirements specification. Clause 5.3.1 of ISO 9000-3 calls for a requirements specification that clearly and unambiguously states what is expected from the product that will be produced. In addition to the functional requirements, the requirements specification should consider such things as product performance, reliability, safety, and security. If the requirements specification is not provided by the purchaser in the contractual arrangement, then it should be produced from a joint effort between the purchaser and the supplier. (Remember that ISO 9000-3 presumes a contractual agreement between a supplier and purchaser where the purchaser either provides, or is actively involved in producing, the product requirements.)

Development plan. Clause 5.4.1 of ISO 9000-3 calls for a development plan that contains plans for the development activities for the project. The development plan should contain the following:

- Definition of the project; states the objectives for the project and references related projects.
- Project organization; describes the team structure that will be used for the project, responsibilities, what subcontractors will be used, and other resources that will be used in the project.
- Development phases; states the various phases that the project will pass through, the required input for each, the expected output from each, the verification procedures to be carried out for each, along with risk assessment and contingency plan for each phase.
- Project management; describes how the project will be managed, how progress will be monitored and measured, organizational responsibilities, work assignments, and interfaces between different groups.
- Development methods and tools; states what tools and techniques, rules, practices, and conventions, etc., will be used during the project.
- Project schedules; shows tasks to be performed, by whom, when, and resources that will be applied to each task during the project.
- Related plans; identifies other plans, such as quality plan, that relate to this development plan.

Verification plan. Clause 5.4.6 of ISO 9000-3 calls for a plan of verification stating how the output from each planned development phase will be verified.

Test plan. Clause 5.7.1 of ISO 9000-3 calls for a document that describes the test plan. The test plan describes the various levels of test activities planned for the project and should include descriptions of the various test activities; test case descriptions with test data and expected results; statement of required test environments, tools, and software; specification of the criteria for completion of the various test activities; and personnel and qualification required.

Maintenance plan. Clause 5.10.2 of ISO 9000-3 calls for a maintenance plan that specifies all the maintenance activities to be carried out by the supplier. The maintenance plan should include definition of the scope of maintenance, description of the initial status of the product upon delivery, identification of any required support organization(s), list of maintenance activities, and description of maintenance records and reports.

Maintenance records. Clause 5.10.6 of ISO 9000-3 calls for maintenance records that provide a record of all maintenance activities in a predefined format. The maintenance records should include all requests for assistance and problem reports, what organization responded to each, priority that was assigned to each, results of the corrective action, and statistical data on failure occurrences and maintenance activities.

Release procedures. Clause 5.10.7 of ISO 9000-3 calls for release procedures that should specify the procedures for incorporating changes into a software product. These release procedures should specify the ground rules for determining when fixes can be made and when a new release of the product is required. It should also specify the frequency and impact to the purchaser that is allowable for product changes. It should specify how the supplier will inform the purchaser of changes to the product.

Configuration management system. Clause 6.1.1 of ISO 9000-3 calls for a configuration management system as a mechanism for identifying, controlling, and tracking the versions of each software item. The configuration management system should accomplish the following:

- Uniquely identify the versions of software items
- Identify the versions of each software item that constitute a version of the product
- Identify the build status of the software product during development
- Control simultaneous updating of a software item
- Provide coordination for the updating of multiple products
- Identify and track changes to software items

Configuration management plan. Clause 6.1.2 of ISO 9000-3 calls for a configuration management plan that specifies how configuration management will be done and what will be under its control during the project. The configuration management plan should include identification of the organizations involved in and responsible for configuration management; description of the configuration management activities; list of configuration management tools, techniques, and methodologies; and specification of when items will be brought under configuration management.

Procedures for identification of software items. Clause 6.1.3.1 of ISO 9000-3 calls for procedures for identifying software items during the development process. These procedures should be applied to ensure that the following can be identified for each version of a software item.

* The functional and technical specifications it supports
* The development tools used during its development that affect the functional and technical specifications it supports
* Interfaces it provides to other software items and hardware
* Documents and computer files related to it

Procedures for tracing software items. Clause 6.1.3.1 of ISO 9000-3 calls for procedures to facilitate traceability of software items. These procedures should allow each version of a software item to be traced back to its origin, including identification of the level of requirements specification supported by the version of the software item.

Procedures for changing software items. Clause 6.1.3.2 of ISO 9000-3 calls for procedures to identify, document, review and authorize any changes to the software items under configuration management. These procedures should include methods to notify those concerned with, or impacted by, the changes.

Status report procedures. Clause 6.1.3.3 of ISO 9000-3 calls for procedures to record, manage and report on the status of software items. These procedures should include status of change requests and status of approved changes.

Considerations
ISO 9000-3 suggests the following list of additional items to consider when designing an ISO 9001 conforming quality system for software development.

4.1.2 Purchaser's management responsibility. The purchaser should cooperate with the supplier by providing timely information and prompt response to issues, and by assigning a qualified representative to deal with the supplier on contractual matters.

4.1.3 Joint reviews. Joint reviews involving the supplier and purchaser should be held to cover software's conformance to specifications, verification results, and acceptance test results.

4.2.1 Quality system. The quality system should be an integrated process throughout the entire life cycle ensuring that quality is built into the product as development progresses. Problem prevention should be emphasized. The quality system should be documented in a systematic and orderly manner.

5.1 Quality system—life-cycle activities—general. Development projects should be organized according to a life cycle model.

5.2 Contract review. When reviewing a contract, in addition to the items listed by ISO 9001, clause 4.3, the following should be taken into consideration: identification and assessment of possible contingencies and risks, adequate protection of proprietary information, the definition of the supplier's responsibility with regards to subcontracted work, agreed-upon terminology by both parties, and the *purchaser's* capability to meet the contractual obligations.

5.3 Purchaser's requirements specification. The supplier should have a complete, unambiguous set of functional requirements that include all aspects necessary to satisfy the purchaser's needs, are testable, are developed in close cooperation with the purchaser (if not provided by the purchaser), are subject to document control or configuration management, and specify all product interfaces. The requirements specification should be complete and approved before the development activities start.

5.5.1 Quality planning. The supplier should prepare a quality plan (a quality plan is described earlier in this appendix). The quality plan should be updated as the project progresses and items concerned with each phase should be completely defined when the phase begins. The quality plan should be formally reviewed and approved. The quality plan can be independent or part of another document.

5.6 Design and implementation. Design and implementation activities should be carried out in a disciplined manner. (The ISO 9000-3 definition of design and implementation is given earlier in this appendix.)

5.6.2 Design. In addition to requirements common to all development phases, consideration should be given to the following:

- Identification of design considerations
- Design methodology
- Use of past design experience
- Subsequent processes (e.g., testing, maintenance, and use)

5.6.3 Implementation. In addition to requirements common to all development phases, consideration should be given to the following:

- Programming rules, languages, naming conventions, coding, and commentary rules
- Implementation methodologies

5.6.4 Reviews. The supplier should hold reviews to ensure the requirements are being met and that the plans are being carried out correctly. Design or implementation should not proceed until known deficiencies are resolved or the risk of proceeding is known. Records of reviews should be maintained.

5.7 Testing and validation. The software supplier should realize that testing may be required at several levels as software parts progress through the process. They should also understand that some of these levels of testing may be combined. A test plan should be produced to describe the required testing. (A test plan is described earlier in this appendix.)

5.7.3 Testing. When testing, consideration should be given to the following:

- Recording test results in a meaningful way
- Noting discovered problems and their possible impact to any other parts and notifying those responsible

- Identifying and retesting areas impacted by any modifications
- Evaluating the adequacy and relevancy of tests
- Considering and documenting the required hardware and software configuration

5.7.4 Validation. The supplier should validate the complete operation of the product under conditions similar to the applications environment prior to making the product available to the purchaser for acceptance testing.

5.7.5 Field testing. When testing under field conditions, the supplier should consider the following:

- What features need to be tested under field conditions
- The specific responsibilities of the supplier and the purchaser for carrying out the test
- Restoration of the user environment

5.8 Acceptance. When the supplier is ready to deliver the product, the purchaser should judge whether the product is acceptable according to previously agreed-upon criteria. The method for handling problems should be agreed upon between the supplier and purchaser. Before carrying out acceptance testing, the supplier should assist the purchaser in identifying the following:

- Time schedule
- Procedures for evaluation
- Software/hardware environments and resources required
- Acceptance criteria

5.9 Replication, delivery, and installation. Replication is the step that is performed before delivery and should consider the following:

- The number of copies to be delivered
- The delivery media
- The required documentation
- Copyright and licensing concerns
- Custody of master and back-up copies, including disaster recovery plans
- How long the supplier is obligated to supply copies

Consideration should be given to verifying the correctness and completeness of the delivered software product.

For installation, consideration should be given to the roles, responsibilities, and obligations of the supplier and purchaser, taking into account the following:

- Schedules, including other than normal working hours
- Access to the purchaser's facilities
- Availability of skilled personnel
- Availability and access to the purchaser's system and equipment
- The need for validation of each installation
- Procedures for approval of installation upon completion

5.10 Maintenance. The requirement for maintenance by the purchaser should be stipulated in the contract. When maintenance is required, the supplier must establish and document the procedures for performing the maintenance activities. (The description of what constitutes maintenance activities is given earlier in this appendix.) The items to be maintained and the period of time for which they will be maintained should be specified in the contract.

The supplier should prepare a maintenance plan. (See earlier section in this appendix for suggested content of the maintenance plan.) When maintenance is provided by the supplier, consider the following:

- Definition, documentation, and agreement on the initial status of the product to be maintained (clause 5.10.3).
- If a separate support organization is necessary, it should be flexible enough to cope with unexpected occurrences of problems. It may also be necessary to identify the facilities and resources needed by this organization for its maintenance activities (clause 5.10.4).
- All maintenance should be recorded in a predefined format. Maintenance records (described earlier in this appendix) should be maintained (clause 5.10.6).

Release procedures (described earlier in this appendix) should be established (clause 5.10.7).

6.1 Configuration management. A configuration management system (described earlier in this appendix) should be used as a mechanism for identifying, controlling, and tracking versions of each software item (clause 6.1.1).

A configuration management plan (described earlier in this appendix) should be developed to describe and plan the configuration management activities (clause 6.1.2).

Procedures for identifying software items (described earlier in this appendix) should be established and implemented (clause 6.1.3.1). Procedures to facilitate traceability of software items (described earlier in this appendix) should be established and implemented (clause 6.1.3.1). Procedures to identify, document, review, and authorize changes to software items (described earlier in this appendix) should be established and implemented (clause 6.1.3.2). Procedures to record, manage, and report on the status of software items (described earlier in this appendix) should be established (clause 6.1.3.3).

6.2 Document control. Document control procedures should be applied to relevant documents including the following:

- Procedural documents describing the quality system
- Project planning documents describing the planning and progress
- Product documents describing a particular software product, including development phase inputs, development phase outputs, verification and validation plans and results, documentation for purchaser and user, maintenance documentation.(clause 6.2.2)

Special attention should be given to appropriate approval, access, distribution, and archiving procedures for computer files (clause 6.2.3).

6.4 Product measurement. Product metrics should be recorded and used to measure and improve the development and delivery process. Some metrics should represent reported field failures and/or defects from the customer's viewpoint. Metrics should be comparable and should be considered for use to accomplish the following:

- Report values on a regular basis
- Identify level of performance
- Take remedial actions
- Establish specific improvement goals

6.4.2 Process measurement. Process metrics should be recorded and used to measure and improve the development and delivery process. Process metrics should reflect the following:

- How well the development process is meeting milestones
- How effective the development process is

6.5 Rules, practices, and conventions. The supplier should establish and use rules, practices, and conventions to make the quality system effective. These should be reviewed and revised as appropriate.

6.6 Tools and techniques. The supplier should identify and use tools and techniques to make the quality system effective. These should be reviewed and improved as required.

6.7.3 Validation of purchased product. If, in order to validate a subcontractor's work, the supplier needs to conduct design and other reviews of the subcontractor's work, this requirement should be included in the subcontract.

6.8 Included software. When software other than that produced by the supplier is included in the final software product, consideration should be given as to how the included software will be maintained. A purchaser-supplied product, intended for inclusion in the final product, that is found to be unsuitable for inclusion should be recorded and reported to the purchaser.

6.9 Training. Additional items to consider as part of appropriate training might include specific tools, techniques, methodologies, and computer resources used in development. It also might be necessary to provide knowledge of the specific field with which the software product is to deal.

Summary

ISO 9000-3 provides excellent guidance and suggestions as to what to implement and what considerations to make when designing a quality system for software development that will conform to ISO 9001. It provides guidance on how to apply ISO 9001 to software development. TickIT auditors will use ISO 9000-3 as their guide when auditing a quality system. To pass an ISO 9000 audit, you do not have to implement all of the items spelled out by ISO 9000-3. You do, however, have to satisfy the requirements set by the ISO 9000 series of standards (i.e., ISO 9001, ISO 9002, or ISO 9003). You should seriously consider the guidance provided by ISO 9000-3, understand the requirements of the standards that each addresses, and then ensure that your quality system adequately satisfies the standards.

Appendix I

TickIT

Recognition that the process for software development and maintenance is different from that of most other types of industrial products has led to industry pressure to devise an ISO 9000 registration scheme for software. A registration scheme for information technology (IT), called TickIT, is being formulated by IT professionals, led by the TickIT project office of the U.K. Department of Trade and Industry and supported by the British Computer Society. The objectives of TickIT are as follows:

1. To ensure that the ISO 9000 series of standards is applied appropriately to software.

2. To ensure consistency of certification within the IT industry,.

3. To enable mutual recognition of registration across the IT industry.

To ensure the ISO 9000 standards are being applied appropriately, the TickIT scheme requires auditors to use the TickIT Guide, which is based on *ISO 9000-3: Guidelines for the application of ISO 9001 to the development, supply and maintenance of software.* The TickIT Guide tends to suggest more of how to implement an ISO 9000 conforming quality system than do the standards, which state what must be done.

Under the TickIT scheme, auditors are required to pass a rigid set of criteria to become TickIT accredited. The combination of the requirement to use the TickIT Guide and the TickIT accreditation procedure should lead to mutual recognition of TickIT registration.

Currently the TickIT initiative is limited to the United Kingdom, although many other countries are considering adopting similar schemes for software. At the time of this printing, however, the TickIT scheme has not gained total acceptance even in the United Kingdom, so its future is still undecided.

TickIT is the juxtaposition of the British word for the American check mark (a tick mark) and IT, the acronym for Information Technology.

Appendix J

Nations Adopting ISO 9000

As of January 1993, the following nations and the European Economic Community had adopted the ISO 9000 standards.

Algeria	Mexico
Argentina	Netherlands
Australia	New Zealand
Austria	Norway
Barbados	Pakistan
Belgium	Philippines
Brazil	Poland
Canada	Portugal
Chile	Romania
China	Russian Federation
Colombia	Singapore
Cuba	South Africa
Cyprus	South Korea
Czechoslovakia	Spain
Denmark	Sweden
Finland	Switzerland
France	Taiwan
Germany	Tanzania
Greece	Thailand
Hungary	Trinidad/Tobago
Iceland	Tunisia
India	Turkey
Ireland	United Kingdom
Israel	United States
Italy	Venezuela
Jamaica	Yugoslavia
Japan	Zimbabwe
Malaysia	

Appendix K

The European Market

There is a European movement—a slow, but determined movement—to unite Europe into a single, powerful global political and economic force to rival the United States and Japan. A first step in reaching these objectives is the buildup of Europe's economic strength. The initial step of the European Community (EC) Commission's plan to remove barriers and obstacles that impede the flow of goods, services, and people among member nations of the community took place on December 31, 1992, when many border restrictions were eased between European countries.

There are currently 12 member countries in the EC representing a market of 340 million people. This, coupled with other European countries that are expressing varying degrees of interest in the EC, could ultimately result in a single European market of over 500 million people.

The EC plan to unite Europe is far-reaching and comprehensive and is generating much enthusiasm in Europe, especially among young people. It involves such things as establishing a European-wide high-speed rail system; an EC television network; a European space station; border controls and barrier removal; joint civilian and military research and development projects; European currency, traveler's checks, and postage stamps; standards for the telecommunications industry; and, ultimately, a European governing body.

The EC plan specifies a comprehensive list of obstacles (financial, technical, and physical categories) in establishing a single market and contains almost 300 remedies, or directives, to address these obstacles. The directives include such issues as complex industrial norms and standards, border formalities, air fares, taxation, subsidies, public procurement, rules governing rights of migrant workers, asylum, and extradition. Many of these directives will require manufacturers of regulated products to have a well-documented, well-implemented quality system. More and more companies are using the ISO 9000 series of standards to document, implement, and demonstrate their quality systems. To help liberalize the flow of goods within the EC, the EC urges manufacturers to maintain, among other things, quality standards by obtaining ISO 9000 certification. This certification provides assurance, with a high level of confidence and without requiring further investigation, to the purchaser that the supplier is capable of consistently providing quality products or service.

The future of the EC movement to unite Europe may be uncertain. However, it is agreed that the liberalizing of the flow of goods, services, and people within the EC is necessary if Europe hopes to compete with the United States and Japan. Regardless of the ultimate outcome of the EC plan, the Common Market remains one of the world's largest trading blocs and a large export market for the United States. It will be easier to do business in Europe with ISO 9000 and may be difficult (perhaps, impossible) without it.

Appendix L

Baldrige, Deming, and ISO

When it comes to quality, W. Edwards Deming, total quality management, and the Malcolm Baldrige National Quality Award are all familiar. Now there is ISO 9000, and one has to wonder how these all relate. Is ISO 9000 consistent and compatible with the others? Are they complementary? Contradictory? Can you build on one to get to the other? Just how do they relate?

These are all good questions. Comparison is difficult because each is unlike the other. The good news, though, is that they tend to complement each other and are *not* contradictory.

Quality based on Deming's teachings is an overall philosophy, an attitude, a way of life, almost a religion, based on the premise that if you make quality your number one objective, properly manage toward that objective, and continually improve, good things will result. Heavy emphasis is placed on how you manage the attainment of quality products and especially on how you manage your employees. The expectation is prosperity through continually improving the quality of your product or service, and that quality is achieved mainly through proper management.

The Malcolm Baldrige National Quality Award is a national award that recognizes U.S. companies that "excel in quality achievement and quality management." It awards the companies that demonstrate the best implementation of quality management. It evaluates how good your quality management system is and how well you've implemented it. The intent of the Baldrige Award is to stimulate the pursuit of quality by recognizing excellence in quality achievement.

ISO 9000 is an international series of standards that suggests a basic structure of what needs to be done during the development of products to ensure a quality product every time. It specifies what needs to be done but does not evaluate how well it is being done. It does, however, require that you be able to demonstrate that what you do is effective. The intent of the ISO 9000 series of standards is to ensure a level of acceptable, consistent quality than can then be further improved. Registration to ISO 9000 is perpetual as long as you maintain and improve your quality system.

Of the three, the ISO 9000 series of standards is the minimum set and, with few exceptions, is a subset of the Baldrige requirements. Baldrige goes into areas not addressed by ISO 9000, but there are no contradictions. Implementing an ISO 9000 conforming quality system would be consistent with Baldrige and would provide an excellent start toward it. ISO 9000 and Baldrige both would provide excellent foundations for Deming; however, the main ingredient to Deming is its management approach, which is not required by either ISO 9000 or Baldrige. If you have not decided on the Deming management approach, moving from ISO 9000 or Baldrige to Deming could require a major management transformation. If Deming quality is your ultimate objective, you should start with the management transformation. Implementing an ISO 9000 conforming quality system, although not necessary, would certainly not hurt the effort and would provide, like with Baldrige, an excellent foundation upon which to build.

Following is a more detailed discussion about each.

Deming

W. Edwards Deming is probably the best known of the quality advocates. He is often referred to as the person who taught the Japanese about quality. Deming quality is an attitude, a philosophy, a way of life. The basic Deming concept is that quality must be first and foremost—almost at the expense of all else. If trade-offs have to be made, they are generally made in favor of quality. And you must work to continuously improve your quality. There is a direct relationship between quality and sales, quality and productivity, quality and profit, quality and competitive position. Good, ever-improving quality will lead to good things, such as increased profits, improved productivity, lower cost, and loyal customers. Loyal customers are the best kind of customers; they, in turn, lead to increased market share, higher profit margins, higher profits, more secure and satisfied work force, and more jobs.

Quality is the responsibility of management. It is up to management (*top management*) to decide that the company is going to be a quality company. It is the responsibility of management to bring about the management transformation that is required for the company to become a quality company. Deming quality is achieved through a number of principles, many of which are unconventional or nontraditional. The following are some of the Deming principles.

- Quality is made in the boardroom. Top management is responsible for establishing quality as the first and foremost objective, and it is management's responsibility to lead to the achievement of the objectives.

- Profits are generated by loyal customers, and loyal customers result from quality products. Loyal customers ultimately lead to a successful bottom line. Attempting to meet profit objectives by cutting costs leads to lower quality and loss of customers and is more likely, in the long run, to be unsuccessful.
- Most defects are caused by the system. Properly motivated workers work hard and usually are not the cause of defects. Most defects are caused by the system over which workers have little or no control. Management is responsible for the system and changes to it.
- Do not rely on mass inspections to achieve quality. Quality is not tested into a product, it is built in. Therefore, it is management of the development process that results in quality, not the mass inspection at the end of the process.
- Workers should feel secure and intrinsically motivated. Workers are not motivated when they are fearful and insecure. Workers perform best when they feel secure, not threatened, and are intrinsically motivated. Workers, who take pride in their workmanship and love their work, are intrinsically motivated. Workers who are offered incentives, given quotas, appraised, rated, ranked, etc., tend to be fearful and threatened. They cannot be creative or innovative. Incentives, quotas, rankings, etc., tend to cause contention and fear and destroy teamwork, which leads to shortcuts and lower quality. Quality is achieved through cooperation not contention. Eliminate merit systems, ranking, rating, incentives, quotas, and management by objectives. Management must not judge workers but work to eliminate obstacles to intrinsic motivation.
- Use vendors committed to quality. Don't choose suppliers on low cost alone. Choose vendors who will provide quality products, especially if their product is part of your product.

The Deming approach relies heavily on management to set objectives (quality, quality, quality); to establish, change, and improve the system; to lead; and to remove barriers to having happy workers. Deming does not outline what needs to be done by the process (i.e., the quality system). He talks more about how to manage quality than what to do to achieve it.

Deming's view is that the process is an integral and important part of developing a quality product and that it can always be, and must continually be, improved. Deming prescribes neither what needs to be done in the process nor how to do it.

Malcolm Baldrige National Quality Award

The Malcolm Baldrige National Quality Award is named after Malcolm Baldrige who served as the U.S. Secretary of Commerce from 1981 until his death in 1987. The award is competitive and is awarded to the best companies based on an evaluation of how well they perform within seven examination categories outlined in the award application. These categories are divided into areas that serve as the basis for the evaluation (and thereby suggest what is important in the development process). The areas of emphasis are management, and management responsibilities; measuring and improving the quality of the product; and knowing your customers, what they want and how satisfied they are. One of the seven categories deals with the effectiveness of your quality system, which is where most of the ISO 9000 standards are focused.

Differences between Baldrige and ISO 9000 generally stem from the fact that Baldrige is competitive and is an award for quality leadership and excellence, and, therefore, its criteria goes beyond the basic requirements of ISO 9000. The following shows some of the differences between Baldrige and ISO 9000. Most of the differences are Baldrige requirements that go beyond the requirements of ISO 9000.

- Product quality. Baldrige evaluates how well you measure and compare the quality of your product. ISO 9000 does not directly address *product* quality.
- Benchmarking. Baldrige requires that you compare the quality of your product and your process against those of your competition and against the best. ISO 9000 has no such requirement. With ISO 9000, however, you may want to measure and compare your product quality as a way to demonstrate the effectiveness of your quality system.
- Quality leadership. Baldrige awards quality leadership. ISO 9000 is aimed at attaining a basic quality system for producing consistent, quality products.
- Quality education and recognition. Baldrige requires that employees be educated in quality and that the company recognize quality contributions of its employees. ISO 9000 has no such requirement.
- Employee well-being. Baldrige requires company concern for the general well-being of its employees, including, for example, safety and education outside of the company. The ISO 9000 requirement is limited to providing adequate employee training for the job.

• Organization involvement. Baldrige requires more organizational involvement (and possibly change) than ISO 9000. It also requires that a company "extends its quality leadership to the external communities . . ." Unlike ISO 9000, though, Baldrige does not specifically call for a quality manager.

Because the Malcolm Baldrige National Quality Award is competitive and awards quality *leadership*, its guidelines go beyond those of ISO 9000. ISO 9000 focuses on the development process and looks at what you do; Baldrige goes beyond the process and evaluates not only what you do, but how well you do it.

ISO 9000

ISO 9000 is an international series of standards that specifies a basic set of requirements for a quality system to provide consistent, acceptable quality products. Its emphasis is on the development process and the management responsibilities associated with the process. It focuses on establishing, documenting, and following a well-controlled, auditable development process that is constantly monitored, reviewed, and improved. It does not deal with management techniques and styles, as Deming does, nor does it concern itself directly with the resultant product, customer satisfaction or benchmarking against competition, as Baldrige does.

Like both Deming and Baldrige, ISO 9000 emphasizes the need for quality objectives and stresses their importance. But, unlike Deming who requires, by definition, that quality be the ultimate objective, ISO 9000 merely requires that ambitious, but attainable, quality objectives be established and understood.

ISO 9000 does not judge how good the quality management system is or how well it has been implemented. It does, however, expect that you are able to demonstrate its effectiveness. Like both Deming and Baldrige, ISO 9000 requires continual improvement to the process.

ISO 9000 provides an excellent foundation for addressing the process areas for Deming and Baldrige, with the understanding that more is required for both.

Bibiliography

Aguayo, Rafael. *Dr. Deming, the American Who Taught the Japanese About Quality*. New York: Lyle Stuart/Carol Publishing Group, 1990.

ANSI/ASQC Q91-1987, Quality Systems—Model for Quality Assurance in Design/Development, Production, Installation, and Servicing. Milwaukee: American Society for Quality Control (ASQC), 1987.

Ayers, Steve J., and Frank S. Patrinostro. *Software Configuration Management*. New York: McGraw-Hill, 1992.

Boehling, Walter H. "Europe 1992: Its Effect on International Standards." *Quality Progress*, June 1990, pp. 29–31.

BS 5750: Quality systems Part 1. Specifications for design/development, production, installation and servicing. London: British Standards Institution (BSI), 1987.

Coyne, Brenden. "1992—Fortress Europe?" *Quality*, June 1990, pp. 17–21.

DeAngeles, Cynthia A. "ICI Advanced Material Implements ISO 9000 Program." *Quality Progress*, November 1991, pp. 49–51.

Durand, Ian G., Donald W. Marquardt, Robert W. Peach, and James C. Pyle. "Updating the ISO 9000 Quality Standards: Responding to Marketplace Needs." *Quality Progress*, July 1993, pp. 23–28.

EN 29001: Quality systems—Model for quality assurance in design/development, production, installation and servicing. Brussels: European Committee for Standardization (CEN), 1987.

Goult, Roderick. "Elements of Good Documentation...How to Structure Documentation Beneficially." *Quality Systems Update*, April 1992, pp. 10–12.

Goult, Roderick. "ISO 9001 Interpretations and Misunderstandings." *Quality Systems Update*, May 1992, pp. 10–13.

Goult, Roderick. "Making it Happen: Handling the Registration Audit." *Quality Systems Update*, July 1992, pp. 12–14.

Guide to Software Quality Management System Construction and Certification using EN 29001 (TickIT Guide). London: British Department of Trade and Industry and the British Computer Society, 1992.

Hillkirk, John. "Europe Upstages Quest for Baldrige Award." *USA Today*, April 27, 1992, p. 1B.

IBM Corporation. *Quality Manual IBM Corporation S/390 Programming Systems (Draft 1)*. February 1992. Photocopy.

IBM Corporation. *Quality Manual IBM Research Triangle Park Networking Laboratory Research Triangle Park, NC*. February 1992. Photocopy.

IBM Corporation. *Translation of Quality Assurance Handbook for IBM Deutschland/Manufacturing*. April 1992. Photocopy.

ISO 9000-3: Quality management and quality assurance standards—Part 3: Guidelines for the application of ISO 9001 to the development, supply and maintenance of software. Geneva: International Organization for Standardization (ISO), 1991.

ISO 9001-1987: Quality systems—Model for quality assurance in design/development, production, installation and servicing. Geneva: International Organization for Standardization (ISO), 1987.

Kalinosky, Ian S. "The Total Quality System—Going Beyond ISO 9000." *Quality Progress*, June 1990, pp. 50–54.

Kendrick, John J. "Certifying Quality Management Systems." *Quality*, August 1990, pp. 38–40.

Krause, Axel. *Inside the New Europe*. New York: Cornelia & Michael Bessie Books/Harper Collins, 1991.

Lofgren, George Q. "Quality System Registration." *Quality Progress*, May 1991, pp. 35–37.

Logica plc. *Quality Manual*. 1987.

Marash, Robert. "Choosing an ISO 9000 Consulting Service." *Quality Systems Update*, May 1992, pp. 29 and 32.

Marquardt, Donald, Jacques Chove, K. E. Jensen, Klaus Petrick, James Pyle, and Donald Strahle."Vision 2000: The Strategy for ISO 9000 Series of Standards in the '90s." *Quality Progress*, May 1991, pp. 25–31.

National Quality Award Consortium, Inc. *1990 Application Guidelines for Malcolm Baldrige National Quality Award.*

Peach, Robert W., ed. *The ISO 9000 Handbook*. Fairfax, Va.: CEEM Information Services, 1992.

Quality Systems Update, comp. *ISO 9000 Registered Company Directory United States*. 1992.

Stern, Gary M. "Quality Time." *Sky Magazine*, May 1992, pp. 22–26.

Stratton, John H. "What Is the Registration Accreditation Board." *Quality Progress*, January 1992, pp. 67–69.

Timbers, Michael J. "ISO 9000 and Europe's Attempts to Mandate Quality." *Quality Digest*, June 1992, pp. 18–30.

Van Nuland, Yves. "The New Common Language for 12 Countries." *Quality Progress*, June 1990, pp. 40–41.

Van Nuland, Yves. "Prerequisites to Implementation." *Quality Progress*, June 1990, pp. 36–39.

Woerner, Susan. "Tools for Quality Management: ISO 9000 and Baldrige Award." *Adhesives Age*, December 1991, pp. 41–43.

Index

151